PLANDEMIC

FEAR IS THE VIRUS.
TRUTH IS THE CURE.

EDITED AND COAUTHORED BY

MIKKI WILLIS

SKYHORSE PUBLISHING

Library of Congress Cataloging-in-Publication Data is available on file.

Cover design by Wendy Saade
Cover photo by Airfox

Print ISBN: 978-1-5107-6554-2

Ebook ISBN: 978-1-5107-6555-9

Printed in the United States of America

Table of Contents

Introduction   v

Prologue   xvii

Chapter One: The Origins   1

Chapter Two: *PLANDEMIC 1*   15

Chapter Three: Debunking the Debunkers   35

Chapter Four: *PLANDEMIC 2*   43

Chapter Five: The Gatekeepers   59

Chapter Six: The Dress Rehearsal   75

Chapter Seven: Gates of Hell   87

Chapter Eight: Fact-Checking the Fact-Checkers   101

Chapter Nine: Ending Where It Began   107

Epilogue   117

Acknowledgments   132

Endnotes   133

In need of context for what you're about to read?
Scan this QR CODE with your cell phone camera to
STAY CONNECTED with the PLANDEMIC TEAM
and to watch our movies for FREE!

Introduction

The hardest thing to explain is the glaringly evident which everybody had decided not to see.

—Ayn Rand

———

The offer to create this book came shortly after the release of *PLANDEMIC: INDOCTORNATION*, while I was spinning within the eye of the media storm. I couldn't do it. I had to pass. I knew that if I created any type of product, the media would obsess on it to sway the masses into believing that my motive was personal gain. Even without a product to sell, they pushed that angle anyway.

The untold truth is, we refused to profit in any way from either *PLANDEMIC* movie. We had nothing to sell except for the truth. We didn't even activate a single paid advertisement. We turned down every opportunity for investment and, instead, raised just enough in donations to cover our expenses. Without the concern of financial return, we were able to give the film away. *PLANDEMIC* was our gift to the people. In the end, it was the people who carried it around the world.

After the first *PLANDEMIC* broke records, I received a multimillion-dollar offer to license the brand. Here's a snippet from a 2021 interview for *Ojai Magazine*,[1] with Reno Rolle, the person who was brokering that deal:

"On the heels of his *PLANDEMIC* project, I was approached by people who specialize in monetizing data because they thought I might be able to get to Mikki," he said. "They suggested emphatically that if they had access to Mikki's database, they would market to that database, and they guaranteed seven figures over the course of one week. I know it sounds incredible, but I've been in direct-response community marketing and these people are very credible and legitimate. Mikki flatly refused, because he was concerned people would think he made *PLAN-DEMIC* for the money."

Why would I, an independent filmmaker, who at the time was living from paycheck to paycheck, walk away from a multimillion-dollar guarantee? It wasn't easy. To be perfectly transparent, there have been moments when I questioned that decision. Prior to the release of *PLANDEMIC*, my family and I had lost our home, work studio, car, and everything we owned in the California Thomas Fire. We escaped with our cell phones, a few hard drives, and the clothes we were wearing.

Our insurance policy lacked in the realm of fire coverage. As a result, we received a settlement that barely touched one-sixteenth of what was lost. So, it's not that we didn't need the money. I just couldn't bring myself to profit from a movie of this nature. Thankfully, my wife fully supported my decision. On that note, 100 percent of my profit shares generated from the sale of this book is going directly to a nonprofit organization that exists to create new schools and higher learning systems for children and young adults.

Unless you've had the experience of being completely censored, silenced, and scrubbed from all forms of digital media, you may not understand what it's like to be gagged in that way. Those who control the global narrative took every measure to ensure that I would not have the ability to defend my good name.

We've always been told there are two sides to every story, but unfortunately, the gatekeepers of free speech have ensured that we only hear one side of the story—*their* side. I began seeking an alternative medium through which I'd have the freedom and reach to set the record straight.

My producer, Erik, suggested that I write a book. I'd had a few offers in the past, but being an author was not on my to-do list. Thanks to Erik's persistence, I finally agreed to allow an investigative journalist to begin interviewing key interviewees and crew members to develop the framework for the book.

A few weeks later, Erik called to tell me he had "good news and bad news." "Hit me with the bad," I said.

He replied, "I just learned that our writer is not on our side. She believes the mainstream narrative and thinks we're crazy."

"Wonderful," I said as I braced for yet another hit piece. "What's the good news?"

Erik answered, "Actually, she *thought* we were crazy. She doesn't think that anymore. Her mind is blown by what she's discovering through her research."

To my amazement, this journalist, who for good reason has chosen to remain anonymous, had the courage and integrity to keep an open mind enough to dig beneath the smears and slander. After reading a rough manuscript, I was inspired to jump in as an author.

That said, I will never take full credit for this book. Highest credit goes to the fine people at Skyhorse Publishing, to Dr. Judy Mikovits, Dr. David Martin, and to my mysterious coauthor, whom I may never meet. I'm equally grateful for my incredible research team, my courageous film crew, and the long list of brave and brilliant doctors and scientists who guided me every step of the way to ensure that the information presented within the *PLANDEMIC* series was bulletproof. Yes, bulletproof.

Despite what critics have said, not one major claim in either *PLAN-DEMIC* movie has been successfully proven inaccurate. In fact, shortly after the release of *INDOCTORNATION*, I offered a $10,000 online challenge to anyone who could prove a single major claim inaccurate. After reposting the challenge globally every other day for six months, I gave up.

No one, not a single critic, fact-checker, or doctor was willing to put their money where their mouth is. Hence our catch phrase: 100 percent censored. Zero percent debunked.

Okay, let's get personal.

Like Forrest Gump, for reasons not yet fully known, I'm often placed at the center of historical moments. These are just a few of the highlights: I was working with inner city youth in South Central Los Angeles when the riots of '92 broke out. If you explore archive news reels, you'll see me standing directly behind Rodney King the moment he uttered those unforgettable words, "Can we all just get along?" That question has haunted me ever since.

Three years later, I had an impromptu dinner with O. J. Simpson just after he was acquitted of murder. I was near the World Trade Center the day the towers went down. After digging for survivors for three days, I was a changed man.

Suddenly, I wanted nothing to do with Hollywood. I made a hard pivot to focus my lens on things that matter. I was filming a PSA for the Bernie Sanders campaign with actress Shailene Woodley the day the Dakota Access Pipeline protest began.

We went straight to the front line, where we remained in service to the people of Standing Rock for over two years. I was filming near the US Capital when it was stormed on January 6th, 2021. More on that later. (SPOILER ALERT: the truth is diametrically opposed to the media's version of why I was there.)

These are just a few of the events I credit for broadening my understanding of fate and faith. I was raised without religion. No church. No Bible. No grace before dinner. Our God was love. Long before I came along, my mother's husband died and left her alone with three small children.

Wounded by the loss of the love of her life, and in fear of losing her welfare assistance, my mother remained alone. When her three kids were in their preteens and teens, a girlfriend encouraged her to get out of the house. They went to a local night club, where she met a handsome sailor with piercing blue eyes. One thing led to another, and my mother ended up pregnant. Barely able to feed and care for her kids, a new baby was the last thing she needed.

Unable to bring herself to get an abortion, she did everything possible to induce a miscarriage. But all the horseback riding she could do wasn't enough to stop me from entering this world. My grandmother was not happy when her daughter gave birth to a bastard child.

In an effort to compensate for my grandmother's indifference, my mother showered me with love. Admittedly, I was a mama's boy. She was my best friend. Mom was diagnosed with cancer when I was in grade school. She was a survivor in more ways than one.

My big brother was diagnosed with AIDS when I was in my teens. He struggled with it for eight years before a new medicine called AZT brought new hope. Though it appeared to all of us that this new miracle drug was doing more harm than good, the man leading the AIDS epidemic, Doctor Anthony Fauci, promised the world that it was our only hope.

My brother's health began to rapidly decline. The gay community had begun warning my brother and my mother that it wasn't the virus, but the medicine that was killing him. But every time they turned the channel, there was America's top doctor surrounded by the world's most beloved celebrities, reassuring the world that his protocol was the only solution.

AZT killed my brother on May 23, 1994. Unable to live with the guilt of not listening to the warnings, my mother invited the return of her cancer. She died just thirty-four days after my brother, on June 26, 1994.

I'd never been through anything like that. I didn't have the tools to process what I was feeling. Like Gump, I ran! I had to get far away from anything that made me remember. I went to the place where orphans hide. I rented the cheapest room available at the Magic Castle Hotel in Hollywood. I wasn't looking for stardom. I came for family. That's where they are, right? At least that's how it appeared to a kid who was raised on sitcoms. With only $1,100 to my name, I had to take the first job I could find.

I worked as a print model for just over one year. It was the first time I got to travel. That part I loved. But I rapidly grew disillusioned by the pretentiousness of the industry and began looking for something more real and meaningful. I became a Hollywood actor. What can I say, I was young and naive.

I began auditioning but just didn't have the skills. My first big break was being invited to study with legendary father of method acting Sanford Meisner. I couldn't believe it. I was so green. Why did he pick me? It was the biggest accomplishment of my life at that point. It gave me a confidence that I'd never had.

Six months into training, Sanford, or "Sandy" as they called him, asked me to remain in the theater as everyone else headed out on break. He had me sit on the edge of the stage, our knees nearly touching as he stared into my eyes. My heart was pounding. I didn't know if he was going to give me the axe or praise my hard work.

Speaking through the tracheotomy hole in his neck, he sucked in a gurgling breath, then told me he wanted to have sex with me. I thought it was an exercise. Surely he wasn't serious. He was so old and frail. He had to be testing me. I smiled calmly, then said, "No thank you."

He didn't blink. I continued, "It's not that I have any judgment. I'm just not . . . gay." Still, not a blink. Filling the uncomfortable silence, I said, "I have no issues with . . . you know . . . gay people. My brother is gay . . . I mean . . . *was* gay . . . He had AIDS."

After a long and intense pause, Sandy finally responded, "OK." With the flip of a hand, he waved me off. I left the theater heavy-headed and confused.

When we all returned from break, Sandy directed me to take the stage. I stood there in silence for a moment. Using his cane, he pulled his dying body to a standing position. He then pointed that furious cane at me and growled, "You don't belong on stage! Get out of here! Go now!"

I developed a reading disorder after that, which made my auditions even worse. That was it for me as an actor. As they say, those who can't do, teach. I took a job as a drama coach to toddlers. Alright, it was more like daycare, but I loved it! I loved working with the kids.

Teaching led to directing one-act theatrical plays. I became the youngest member of The Playwrights Kitchen Ensemble, where I was mentored by legends of stage and screen. PKE was the brainchild of Hollywood mogul, Steve Tisch, who produced *Forrest Gump*, ironically, as well as many other iconic classics.

Fueled by the love of theater, I went on to build my own playhouse in North Hollywood, where I began honing my writing and directing skills. Eager to get behind a camera, I raised a few thousand dollars to create my first microbudget mockumentary called *Shoeshine Boys*. To my surprise, that little movie went on to become an underground hit, winning top honors at various film festivals.

In 2001, I flew to New York to meet with a potential distributor. I was living the dream. Not only was I in negotiations for distribution of my first movie, but I was making thousands of dollars a day as a fashion photographer and as a director of Spanish-language music videos. But all that changed on September 11, 2001.

I was sleeping on a friend's sofa in midtown the morning the planes hit. My buddy and I went directly to the scene, where we remained for three days while digging for survivors. This was my wake-up call.

While standing on the rubble of the World Trade Center, looking down at scattered body parts, something happened to me. Something mystical. I could feel the eyes of the world focused on that very spot. The planet was shrinking. Nothing was far away. I could literally feel the presence of every living being. I felt our collective pain. Our fear. Our desire to live and love.

The moment was shattered by an announcement. Every rescue worker was ordered to turn off their machines, stop, and listen. We were told that the dust we were breathing was laced with extremely deadly toxins. Anyone without a proper respirator was invited to leave the area. Not one man or woman walked away.

The announcer made it painfully clear: "What you are breathing will eventually kill you!"

The workers looked around to see if anyone was going to heed that advice. The sound of heavy equipment fired back up, and everyone went back to work. Not a single person left. I stood there, eyes flooding, and said to myself, "This is who we are. This is who we are."

Everyone was willing to risk their life at the fading chance of saving one stranger. I'd never witnessed such selflessness. After that, I began to see people in a new and brighter light. Suddenly, all of my material goals felt trivial. I couldn't imagine returning to Hollywood to do the work I was doing before. How could I direct another commercial to sell a product that poisons our people and our planet? How could I direct another music video that glorified the ego and used women as props? My career was over.

I returned to California, put everything I owned in storage, then moved into a friend's guest cottage in Napa Valley. Still, the gravity of Hollywood kept sucking me back into the machine.

I was offered to write and direct the sequel to the '60s classic *Easy Rider*. As a former motocross racer and fan of anything on two wheels, it was an offer I could not refuse. I signed the deal, wrote the script, then just before the movie went into production, I quit. I just couldn't do it. Not only because it was a bad idea to begin with, but being back in the maze that I had recently escaped was simply something I wasn't willing to endure again.

I walked away from a $400,000 paycheck and never looked back. If I was going to continue to work as a filmmaker, it was going to be on my terms. My newfound clarity and commitment to living truthfully prepared me to meet the love of my life. Nadia and I fell in love in 2003 and were married in 2009. Together, we created the Elevate Film Festival, which became the world's largest single-screen film event. After a three-year tour, we decided to morph the festival into a film production company dedicated to elevating human consciousness.

Nadia went into labor in July 2011. Our home birth plan was scrapped due to severe complications. We rushed to the hospital, where Nadia would undergo an emergency C-section. After much effort, a tiny purple body was pulled from her belly. No crying. No breathing. The doctors placed our lifeless son on a cold machine and began working frantically to pump life into him. Thankfully, Nadia was unable to see what I could see. She asked, "Is everything okay?"

That was the only time I've ever lied to my wife: "Yes, my love. Everything's fine." She smiled, her beautiful dimples popping out. I forced a smile, then returned my eyes to the drama across the room. The machine was now making a sound that I will never forget. The sound of death. The look on the nurse's face said everything. She did her best to give me a reassuring smile, then used her body to block my view as doctors shoved suction devices down my baby's throat.

I closed my eyes and began to pray. Without much experience, I wasn't sure who to address my prayer to. Father? Mother? God? Buddha? Krishna? Christ? With so much at stake, I prayed to all of them. I prayed hard. It wasn't working. I began to beg. I made promises to anyone and anything that might be listening. Nadia asked, "Are you sure everything is okay?" I couldn't lie again. I cradled her face in my hands, then told her the truth through my eyes. Her voice broke as she asked, "What's wrong? Honey, what's wrong?" Once again, I closed my eyes. This time, I made an offer.

I said, "Please God, let that baby breathe and I vow to you right here and now that I will dedicate the rest of my life to this child and all of your children." At that exact moment, a tiny voice cried out. One of the doctors yelled, "That's what we want to hear! That's what we want to hear, little guy!" The machine stopped making that awful sound. The nurse, clearly emotional, smiled big and said, "That's your baby."

Nadia echoed, "That's our baby?" I nodded and said, "That's our baby." We cried together. That was the day I learned to pray.

As I write this, I'm fully aware of this risk I'm taking to share stories so personal. I'm aware of the distrust and cynicism that's currently plaguing our nation and our world. I anticipate that some readers will totally miss the point and my intention for sharing. In no way do I see myself as any kind of hero or martyr. I'm not looking for sympathy or praise. I'm not interested in winning anyone's acceptance. I chose to share these stories because I want you to know the truth. I want you to know the real reasons I made the leap from a lucrative and safe career to produce a movie like *PLANDEMIC*. Contrary to popular media narratives, I have no interest in being famous. Why would anyone in this age of cancel culture shoot for something so fragile and toxic? Furthermore, if money were my goal, I would've taken that multimillion-dollar offer and run.

The corporate media would also have you believe that I am a far-right radical of sorts, despite the fact that up until recently, I was as far left as one could be, without falling over the edge. Now that I've been behind the curtain of politics for the last few years, I currently identify with neither of the two parties. Seeing firsthand the trappings of identity politics, I've learned to vote for policies over personalities.

I am also not a "QAnon follower." In fact, to date I've not seen a single "Q drop," as they call it. The reason is simple. As a professional researcher, I only pay attention to information that can be validated through verified sources. That said, I hold no judgment for anyone in that movement. The few Q followers I've had the pleasure of meeting were genuinely good people. That's what matters to me.

With all the effort to dehumanize and divide us, I refuse to participate in that losing game. Through my work as an interviewer, I've learned the importance of listening. We all have a story in us. To listen to one another's stories is to reconnect as humans. Connection is vital. May the stories within this book leave you more connected with yourself, your loved ones, and all of humanity.

—Mikki Willis

Prologue

Our lives begin to end the day we become silent about things that matter.

—Martin Luther King Jr.

New York
April 2021

———

This is a book that never should have been written.

To start, much of what is described in the following pages was entirely preventable. As you read it, you'll recognize the crucial junctures where a different decision could have changed the path of human history and saved hundreds of thousands of lives.

Really, though, this book never should have been written because I never should have written it. In my nearly four decades on this planet, I've hardly ever had cause to question the medical establishment.

I followed the recommendations of the Food and Drug Administration (FDA) and rolled my eyes at "anti-vaxxers." As it became more of a political statement to do so, I could say without hesitation that I believed science (and women, for that matter). I'd never, *ever* voted Republican. In short, when the first *PLANDEMIC* video began to make its way into my social media feed, I averted my eyes and kept on scrolling. I was not sympathetic to its worldview. At least, that's what I thought.

For some of you reading, that might be reason enough to ignore the rest of what I have to say. The world has become so politicized and divided, certain words and phrases act as triggers that slam shut the door to any kind of open conversation or critical inquiry. *Vaccine* is one of them. *Democrat* or *Republican*, too. Yes We Can. Make America Great Again. Guns. Science. Black Lives Matter. Believe All Women. Blue Lives Matter. Not My President. Crooked. Rigged. Stolen. Liar. Is there anyone left reading this who hasn't felt some kind of reaction by now?

Despite the discord in our nation, however, underneath all of the words that we use to try to make sense of our world, there is still a bedrock of unassailable facts. (And I don't mean *alternative* ones.) As a lifelong investigative journalist, it's been my passion and my duty to uncover them—especially when someone is invested in keeping them out of sight.

Because "journalist," "news," and "facts" can be trigger terms these days, you should know that I've never been a devoted member of what one might call "mainstream media" on either side of the aisle. My books are available at your local store, and you've probably seen my byline on the front page of your paper. Otherwise, I've managed to stay relatively independent. Beholden to no one at this point in my career, my latest investigations have remained largely unclouded by the pressures of money, politics, and corporate powers-that-be. My motto is the old George Orwell yarn: "Journalism is printing what someone else does not want published; everything else is public relations."

For that reason, my journalistic spidey sense began to tingle as the pandemic rolled on throughout 2020. The instances of obvious doublespeak, backtracking, and about-faces when it came to the "truth" were piling up. Knowing—in some cases personally—the reporters at other media outlets, I was painfully aware that they were mostly too lazy to do anything other than regurgitate whatever they were seeing on Twitter or the newswires. So, I started to do my own research in the hopes of understanding why the world seemed to be crumbling around us.

PLANDEMIC inevitably was part of my research—initially, just as a cultural artifact that I thought it would be easy to disprove. I thought it would be the embodiment of the antitruth, antiscience, highly politicized reaction to the pandemic. As I went down the rabbit hole, though, I realized that wasn't the case. I struggled to find anything about which the *PLANDEMIC* team had been flat-out wrong. In reading other critics' takedowns, I read between the lines and saw that while they weren't happy with its message, they didn't ever provide any evidence to suggest that the claims in the film were lies.

I was so curious: How did the filmmakers behind *PLANDEMIC* (both Part One and the second release, *PLANDEMIC: INDOCTORNATION*) create a movie that was both so explosively controversial and so doggedly straightforward? Why did it become such a cultural phenomenon, and what does that say about the human experience of the COVID-19 pandemic? I reached out to them to find out.

If you've picked this book up, you probably think you know the answer, and you probably have an opinion of the film itself—even if you've never watched it. Either way, and no matter what you think, my request to everyone reading this is the same: please try to keep an open mind and remain aware of when that door in your mind is starting to swing shut because you're triggered.

COVID-19 has been the most consequential experience of most of our lives. We owe it to ourselves, to the millions who've died from it, and to generations to come, to try and figure out what happened—and if it really had to happen that way. My opinion? It didn't.

With lockdowns rolling back and case rates dropping, it may be tempting to push forward and forget that this whole ordeal happened. Unless we're willing to confront the truth of what we've all experienced, the horrors of the last year won't be behind us. They'll only be beginning.

Am I confident that we can learn from this massive human tragedy and move into a better era? I'm not so sure. That's why you won't find my name on the cover or inside the pages of this book. It's not because I'm not willing to stand behind what I've reported and written. I do, and I do so proudly. The reason that I am writing anonymously—at least, for this edition—is that I'm not willing to sacrifice my safety, my career, and my family over other people's projections.

There are people who will read this book soberly and judge it on the merit of its factual evidence and arguments. There are other people, however, who are probably already writing up their Amazon review of the book now after reading just a few pages. One-star or five-star, it doesn't matter. I'm not willing to put myself out there to be judged by people who are judging me based on something other than the facts.

Why bother writing the book at all then? I'm not ready to give up on the power of one human sharing a story with another. It's how this great international society started, and it's ultimately what we'll have to come back to if we have any shot of healing the divides in our nation and our world.

So, as you read through this book, I beg you to listen: with your heart and with your mind. If you walk away feeling exactly the same way you do right now about COVID-19, and you feel like you haven't learned anything new or changed your perspective one bit, then by all means write that one-star review. However, if you find yourself changed by the time you read the final page, please don't keep it to yourself. Tell this story to someone else. It's a story of tragedy, conspiracy, and death, but also of a lot of hope, joy, and optimism for the possibilities of the human experience. That story starts now.

CHAPTER ONE

The Origins

*I am a firm believer in the people. If given the truth,
they can be depended upon to meet any national crisis.
The great point is to bring them the real facts.*

—Abraham Lincoln

**Xiaohongsan, China
The Wuhan Institute of Virology
December 2019**

Researchers in full hazard gear moved quietly beneath the fluorescent lights of the giant concrete building that housed the Wuhan Institute of Virology. White space suits. Giant green gloves. White plastic boots like a child would wear for puddle jumping. Overall, the effect would have been comical . . . if the lab hadn't been filled with deadly pathogens.

The researchers were used to the air of danger that pervaded the facility. Just one of the invisible particles that they handled every day could wipe out an entire city. Incidentally, there was a city of eleven million people surrounding them. The responsibility was heavy—and to some, they weren't up to the job.

The world had heard of SARS in the early 2000s. In 2012, there was the report of another coronavirus outbreak (this one called MERS, or Middle East Respiratory Syndrome). But while the world was distracted by a virus associated with camels, few knew that a potentially deadly SARS strain had been detected in China in 2013. This pathogen—code-named WIV1 (and named for the Wuhan Institute of Virology)—attracted little attention except from the US and Chinese researchers funded by the National Institute for Allergy and Infectious Diseases (NIAID) and Anthony Fauci. By 2015, Dr. Ralph Baric of the University of North Carolina and Dr. Zhengli Shi of Wuhan had performed research that had concluded ominously that the Wuhan coronavirus was "poised for human emergence."

If it was going to happen anywhere, Wuhan seemed a likely place. As early as 2016, American researchers found that China was suffering from a "shortage of officials, experts, and scientists who specialize in laboratory biosafety." The greatest concern was that lab researchers who were accidentally infected through lax safety protocol could then inadvertently spread rare diseases throughout their community. Still, that nation's leaders seemed intent on pressing forward with ever more biomedical research.

When the Wuhan Institute of Virology first officially opened in 2017, scientists around the world warned that operations at the $44 million lab were a recipe for disaster. The SARS virus had escaped from a major lab in Beijing multiple times, and despite the government's promises of unparalleled safety in Wuhan, the risk to the rest of the world was obvious: Wuhan would be home to more than

1,500 virus strains. Could a deadly virus escape right under the noses of the researchers?

Early indications were not good. According to the US State Department, American Embassy officials in Beijing recorded at least two official warnings about the lab's insufficient safety measures in early 2018. However, it wasn't just the Americans raising alarm. Although Chinese media have historically been slow to admit the failure of government projects, even the propagandistic national press reported that security inspections had discovered several incidents and accidents at the lab in Wuhan.

One security review in particular concluded that the lab had failed to meet national standards in several categories, especially as it concerned the handling of the bats that had been captured for study of the coronaviruses they carry. Researchers admitted to investigators that there had been bat attacks that left them splattered with bat blood or bat urine on their skin. That kind of bat-to-human contact was exactly the kind of interaction that the outside world feared. Even a less-noticeable bat interaction with another lab animal could cause a chain reaction of infection—one that could potentially cripple the entire world.

Still, in the face of a moratorium making much of that kind of research off-limits in the US, the National Institutes of Health (NIH) continued to funnel money to Wuhan to study coronaviruses in bats. More alarming, the study also funded research into mechanisms that would make bat-derived coronavirus deadlier to humans. The NIH grants to the EcoHealth Alliance, which funded research in Wuhan, would continue up through April 2020. This wasn't random.

In 1999, the National Institute for Allergy and Infectious Diseases (NIAID) under the leadership of Dr. Anthony Fauci began funding research into recombinant coronaviruses. Their specific aim was to create "infectious, replication defective, coronavirus." In short, they sought to use coronavirus as a technology that could infect humans without a high risk of transmission. This work, done at the University of North Carolina Chapel Hill, resulted in US Patent 7,279,327:

"Methods for Producing Recombinant Coronavirus," filed in 2002 before Severe Acute Respiratory Syndrome (SARS) existed.

Research into coronaviruses had been heavily funded as a means to harness the highly manipulatable virus for several potential applications in both medicine and bioterrorism. In the United States, the Centers for Disease Control and Prevention (CDC) jumped to file patents on the gene sequence of the coronavirus itself. Although naturally occurring phenomena cannot be patented, any scientific procedure used to study one *can*. Patenting coronavirus meant that the CDC could control future study—and future vaccines. Based on the number of coronavirus patents that arose in the late 1990s, they foresaw a busy—and potentially profitable—future for that viral family.

All that was likely swirling in the mind of lab director Wang Yanyi in December 2019. An unexplained wildfire of pneumonia had been spreading across the Wuhan metropolitan area for weeks, and doctors had traced it all back to a coronavirus. Yanyi and his team had been tasked with finding out if this coronavirus was a long-buried strain that had resurfaced, or if it could be something new—and therefore much more dangerous.

The results of their initial research were disturbing: This virus did have 96 percent genetic similarity to a strain of coronavirus that had been isolated from bats nearly twenty years before. However, beyond that, it appeared to be something entirely novel.

Samples of the virus reportedly collected from patients arrived in Wuhan on December 30, 2019, and the lab's scientists had reported the viral genome sequence by January 2, 2020. The news of the novel coronavirus was reported to the World Health Organization (WHO) on January 11. According to a Stat News Report article released on January 11, Chinese national media reported the first official death from the virus.[1]

On July 9, 2021, Organic Consumers Association reported that Dr. Ralph Baric, the NIAID, and Moderna entered into a Material Transfer Agreement to start making a new coronavirus vaccine on December 12, weeks before the "pathogen" was isolated.[2]

Lab director Yanyi and the rest of the world now knew what they were dealing with. But where did it come from, and how did it start infecting humans? That was probably a less important question than this one: Was it too late to stop it?

Ojai, California

The sleepy mountain town of Ojai, California, couldn't be farther away from a Chinese coronavirus research lab. About an hour and a half up the road from Los Angeles, Ojai is far removed as well from the hustle and bustle of Hollywood. Getting there involves a slow and steady journey up a winding mountain road, a drive that requires a literal change of pace. As you motor through the natural arches of centuries-old trees, sparkling lakes pop out from behind the bends. Charming farmhouses are nestled in the greenery. Then, suddenly, there is a small town seemingly dropped into the forest out of nowhere.

Spanish-style adobe buildings with wooden signs line the one narrow thoroughfare of commerce in the city. Vegan restaurants live happily alongside coffee shops, tax preparation firms, lawyers, and design studios. Tucked away on a small side street, at the top floor of a dark and nondescript commercial building, was the office of Elevate Productions.

Elevate was the brainchild of Mikki Willis, his wife, producer Nadia Salamanca, and an international team of collaborators. The road to its creation was a rocky one for Mikki, who experienced the deaths of his brother and mother just a few years before coincidentally finding himself at the World Trade Center on 9/11.

Although his experiences were ones that might have turned another man bitter, Mikki ultimately found a deep sense of connection and meaning in the experiences. Frustrated that the news media did not seem interested in telling the positive stories of humans working together in 9/11 rescue efforts—focused as they were on the tales of tragedy and terror—Mikki abandoned a promising career as a hotshot Hollywood director to tell stories about the good in life—and to encourage others to do the same.

"Before my experience at the World Trade Center, I was driven to obtain all the material fetishes we've been wired to see as symbols of success. All that stuff they strive for in Hollywood," Mikki told me in an interview. "But there I was, standing on the rubble of what was an international symbol of power just moments before . . . watching exotic cars being flipped and crushed by rescue vehicles, while body parts lay scattered around me. . . . Suddenly my life goals felt insignificant."

He continued, "It was a snap to grid moment for me. I could no longer do the work I was doing before. I was living someone else's dream. If I was going to remain in 'the business,' I'd have to be involved in something more meaningful."

In 2005, that declaration took the shape of what would come to be called the Elevate Film Festival. "It was more of a guerrilla filmmaking competition than a traditional film festival," Mikki explained. "The object of the game was to challenge filmmakers from around the world to produce a short film in a micro amount of time. We gave each film-maker a small budget, then sent them out into the world to find stories that would lift the human spirit.

"Tired of all the negative news and depressing narratives, our goal was to inspire artists and storytellers to focus on the upside of humanity—all of the innovators, heroes, and great things happening around the world."

What started as a small gathering in a local yoga studio rapidly attracted audiences of up to 6,000 people, filling arenas such as L.A.'s Nokia Theater. As director of the festival, Mikki was tasked with developing each film assignment. One such film assignment was a documentary about urban farmers. "Most of the farmers were immigrants—some legal, some not— and they had developed a beautiful garden, right in the middle of the most industrial areas of South Central Los Angeles. They turned a concrete jungle it into an incredible oasis where they were growing and selling organic food to benefit the entire community," he explained.

Just as the gardens were in full bloom, the owner of the land, a real estate mogul, decided to sell the entire block. "We created a short film titled *South Central Farmers*, then blasted it out to help raise awareness. Overnight, media and thousands of people showed up to stand in solidarity with the farmers and families who relied on the gardens to survive. It was my first experience of producing a piece of media that caused people to take right action. It lit a fire in me!" Mikki explained.

"I began to pay attention to things that I had always avoided," he continued. "Like politics. Though I was deep into my thirties, I had never voted. Barack Obama was the first candidate to inspire me enough to take that leap. I was so enamored by his hypnotic presence that I teared up the night he was sworn in. I was certain that this beautiful family man would deliver on his promise of 'Hope and Change.' By the end of his first term, it was clear that he was like all the rest. A politician. I didn't think I'd ever vote again."

Then, along came Bernie Sanders. "People who I love and trust swore that he was different," Mikki said. They sent links to videos of Bernie dating back decades. His message was consistent. He took me back to my childhood. He spoke about single mothers and how those on the bottom need to be lifted up. I remember thinking, 'I wish we had him when I was a child!'"

Ever intent on sharing solutions with his friends and fellow activists, Mikki began to promote Sanders online and became active in various Internet groups related to the campaign. When he heard that Sanders would be making a campaign stop in Ventura, CA—a short drive from Ojai—Mikki set out to attend his first political rally. He wouldn't be attending just as an observer, though. He intended to film the proceedings. After asking for and receiving permission from the Sanders camp, Mikki showed up on the big day with his camera in tow.

Prior to the rally, he filmed a press conference hosted by celebrities. "An old RV pulls up and out steps Rosario Dawson and Shailene Woodley," he recalled. "I was behind my camera when Rosario looked directly at me. Her eyes got big, and she mouthed the words, 'Oh my

God,' then waved to me. I looked over my shoulder to see who she was waving to. There was no one behind me."

"She came right up to me and said, 'I love you,' then gave me a big bear hug. I figured she had me confused with someone else, but I wasn't about to reject that hug. I said, 'I love you too!' And I meant it. I had always admired her onscreen, and I'd seen her on video speaking at Bernie rallies. I just wished that I was whoever she thought I was."

As it turned out, Dawson knew *exactly* who Mikki was. He had been making home movies and posting them on his Facebook page. One of those videos even reached 100 million views—and one of them was Dawson.

In the one-minute clip, Mikki is seen in the car with his sons, Azai and Zuri. Speaking directly to his cell-phone camera, Mikki explains that Azai had received two of the same birthday gifts at his party, so the duo went to the toy store to exchange one of them. Azai's choice? A doll made in the likeness of "Ariel" from *The Little Mermaid*.

"How do you think a dad feels when his son wants to get this?" Mikki asked in the video, posted on YouTube on August 23, 2015. Smiling big in the background, Azai chimes in, "Yeah!" Mikki responds, "Yeahh! I let my boys choose their life. . . . We say, 'Yeah! Choose it. Choose your expression. Choose what you're into. Choose your sexuality. Choose whatever.' And you have my promise, both of you, as we sit in this car—this hot car in this parking lot—you have my promise forever to love you and accept you no matter what life you choose."

Mikki had been recording sweet moments with his sons almost since their birth, but there was something special about that clip. The video went around the world, and Mikki was invited on major TV shows to talk about his favorite subject: fatherhood.

He soon learned, however, that his message was being misconstrued. It was the line "choose your sexuality" that was at the center of the brewing storm. "I didn't expect my sons, who were only two and four at the time, to understand what those words meant. It was a message intended to reach them once they were mature enough," Mikki explained. "I simply wanted my boys to know that the world and their personal choices could never dilute my love for them. What I wasn't aware of at that moment in time was the emerging agenda to erase gender identities."

"Let me make this point crystal clear," he continued. "I am about personal freedom. It's not my job to judge others for the way they live their lives, so long as they are not doing harm to others, or our environment. How can I expect to live free if I don't grant that right to others? Be who you were born to be. If your choice is to live as a straight person—do your thing. A gay person—cool. Gender fluid— you do you. But let us be wise enough to recognize the potential hazards of allowing any new ideology the power to erase our nature. After all, in my humble opinion, it's our separation from Nature that's at the root of every issue we're currently dealing with.

"To me, the term 'sexuality' refers to the style in which we choose to express our uniqueness as beings capable of procreation. My sons are boys," he said. "One day they will be men. My job is to guide them to become the best men they can possibly be. If for any reason one or both of them choose to express characteristics that fall outside of what might be traditionally defined as masculine, I will fully love and support them. Again, it's about freedom. Freedom to choose. As a former rebellious young person, and now a parent, I'm clear that the more I attempt to mold my boys into *my* vision for them, the more they will push in directions that they may not otherwise choose for themselves. My job is to always be there for them, and out of the way at the same time."

Mikki's home movies struck a chord with a generation hungry for healthy father role models. Dawson had been one of the people tracking Mikki's videos. "She said, 'I share your videos with everybody,'" he said. "I remember thinking, *Oh, wow. She really does know me. This is amazing!*"

From then on, Dawson took Mikki under her wing, walking him around the event and making introductions. She also connected him with actress Shailene Woodley, who bonded with him over his activist roots. They'd only just met, and they were already fast friends and colleagues.

"Shailene and Rosario said, 'We're doing a US tour. Come with us!'" Mikki recalled. "I dialed my wife, Nadia, to see how she felt about me going on the road. Unsurprisingly, she said, 'Oh my god. Do it.' My wife is amazing. I rushed home, packed up, kissed my family good-bye, then hit the road."

Although he was never officially hired by the campaign, Mikki had carte blanche because of his association with the campaign's most prominent supporters. "I'm backstage. I'm onstage. I'm wherever I want to be," he said. "I wasn't officially hired by the campaign, nor was I ever offered pay. At the chance that this gruff old guy could bring balance to our topsy-turvy country, I was happy to pay my own way and to work for free. I created a series of short promotional videos to boost support of the blooming grassroots movement."

However, not everyone was as enthused. Mikki began to receive messages from friends who were concerned to see him latching his cart to the Sanders train. A few of those warnings were from people who had immigrated from socialist countries. One was from Sanders's home state of Vermont and knew Bernie and his family personally. Mikki believed in the vision so deeply, it was hard for him to even consider the warnings.

"I was not open for debate," he told me. "In so many words, I told my friends, 'I really appreciate your time and effort, but I've encountered Bernie, his wife, and even his grandchildren, and I really like the man. So thank you, but this doesn't change anything for me.'"

One friend went so far as to claim that Sanders—an outspoken critic of Hillary Clinton—would end up endorsing her. At the time, Mikki found that utterly unthinkable.

"My friend said, 'Bernie will eventually endorse Hillary Clinton,'" Mikki remembered. "That's when I said, 'Alright. Now I *know* I shouldn't listen to you because that's ludicrous. There's no way in hell. This man has spent most of his career fighting against people like Hillary Clinton and corrupt organizations like the DNC. You're wrong.'"

"Though I was fairly new to the world of politics, I was well versed in the history of Bill and Hillary," Mikki explained. "My mother was from Arkansas. Her brothers—my uncles—had direct experiences with the Clintons. I heard the legends of organized crime and corruption since I was a young boy. As an adult, I looked into those claims and found mountains of supportive evidence. As much as I'd love to see our nation in the hands of a good woman one day, Hillary Clinton was not the one."

The moment of truth came in late July 2016, when Mikki and the rest of the Sanders tour stopped in Philadelphia for the Democratic National Convention. Sure enough, Clinton was declared the nominee. Sanders conceded and forfeited his campaign contributions, later signing a pledge of loyalty to the DNC.

"I was with a large group of loyal Bernie supporters when he conceded to Clinton," Mikki said. "No one could believe what had just happened. We were crushed. I booked a red-eye that night and went straight home."

Any shred of hope that politicians could help change the world was destroyed for Mikki that day. Still, he held onto the belief that regular everyday people could create meaningful transformations—and that film could be a powerful way to showcase them.

Through his friendship with Shailene Woodley, Mikki got drawn into the story of the North Dakota protests against the Dakota Access Pipeline, flying north to capture the protests on film. There, he also became an ally of the Lakota People's Law Project and began creating short films with tribal elders to bring awareness to the situation in North Dakota, and the legal plight of protestors who had been arrested.

"We made videos for each 'Water Protector' that was facing bogus felony charges," Mikki told me. "We had a 100-percent success rate. Charges were either reduced to misdemeanors or dropped altogether. Experiencing the power and potential of filmed media and honest storytelling to bring justice for innocent people turned up that fire in me."

Fulfilling as it was to make a difference in the lives of the Lakota People, it was also the beginning of a new era for Mikki—and for those who would see his films. "That is what turned me on to this area of my work that I now refer to as forensic filmmaking," he explained.

In January 2019, Mikki thought he had discovered another underdog in Nathan Phillips, a Native American activist who got into a standoff with teens from Covington Catholic High School during a day of protests in Washington, D.C. A video clip showing high school junior Nicholas Sandmann facing off with Phillips had gone viral, and Mikki was ready and willing to pile on to the "canceling" of Sandmann and his classmates that was happening online.

"I set out to make a video to further support the Native Americans that were impacted by what I thought was a horrific hate crime," he said. Mikki tasked his team with gathering every video clip of the incident that was captured that day. What he found shocked him.

"A couple of days into watching all of the footage, it became clear to me and my team that the kids were set up," he said. "They had never surrounded Native American elders. They never chanted, 'Build the wall,' like all the headlines claimed. They didn't even speak a single derogatory word. The boys were targeted for wearing red Make America Great Again hats, which they bought from a street vendor earlier that day, just so they could identify their fellow classmates while exploring Washington during a field trip together."

Those red hats, Mikki believed, made the boys political targets. "In the eyes of the media and those infected by it, those boys represented everything wrong with America. They were male. They were white. They were Catholic. Worst of all, they were seen as mascots for Donald Trump," he said. "That makes them subhuman."

Mikki was faced with a major conundrum. He thought, "I've never done anything that could be perceived to support the political right, Republican, or any part of that world. If we tell the truth, we will be thrown into that dreaded basket of the deplorables, and that's the most dangerous place you can be right now. But these are 15-year-old boys."

"Not only were they minors," he continued. "But also, they were clearly innocent of the crimes they were being publicly persecuted for. I had their personal cell-phone videos that allowed us to isolate what the students were saying to one another and to the mob that surrounded them. For a large group of teenaged boys, they were extremely well behaved. The only potentially distasteful moment we could find was when the boys began doing the 'tomahawk chop,' a hand gesture commonly used by fans of sports teams such as the Florida State Seminoles, the Atlanta Braves, and the Kansas City Chiefs. Were the boys being disrespectful, or were they simply trying their best to relate with a culture they had only experienced through television?

"According to a parent who was present as a chaperone that day, the boys had no idea that such a widely accepted hand gesture could be seen as symbol of disrespect. She insisted that the boys were loving the beat of the native drum and were only trying to bridge the communication gap," he continued. "My team and I were faced with a very tough decision: scrap the project or cross a line that we may never be able to return from. We chose to cross that line. As a father of two young boys, I just could not bring myself to look the other way.

"So, the man who had devoted weeks and months of his life to the Bernie Sanders campaign, who had stood alongside protestors at Standing Rock, and been on the side of progressive activism for years, put out a video that told a different kind of story. At least, it would have seemed like a departure to many who knew him. Still, to Mikki, it was the same kind of story: one about underdogs who deserved to have the truth of their story told in the face of a chorus of much louder voices.

"The fifteen-minute video went viral," Mikki explained. "People swiftly discovered that this Nathan Philips had never fought in Vietnam as he claimed on camera more than once. It was also revealed that he had done this sort of thing before and has an MO of crying victim, then raising thousands of dollars through crowdfunding."

"As expected, the haters came at us," Mikki continued, "accusing us of siding with racists, white supremacists, the colonizers, Nazis, etc. As I dug deeper to understand how so many people were willing to throw innocents kids under the bus, I discovered that the vast majority of the complaints came from white people.

"Several of my Native friends reached out to thank me for helping to draw a distinction between them and what was clearly another hate-hoax. Few within my political party saw it that way. Overnight, I went from being a hero of the left to their latest villain.

"It was shocking to experience firsthand how fragile so many of my relationships were," he continued. "People who were never shy in expressing their love and appreciation for me were suddenly trying to destroy me. This was the first time that I received death threats. But not the last. Not one person was interested in learning what led me to create that video. No one wanted to see any evidence that conflicted with the mainstream narrative. Though it was all right there on video, they could not see beyond those red hats."

It was that experience that led Mikki to begin producing a feature-length documentary called *The Narrative*—an investigation into the ways that American media distorts the truth and leverages our differences to divide us. Speaking to whistleblowers and counterculture activists around the world, Mikki soon saw a common thread emerging: the world was headed for a major disaster.

"I was in the process of interviewing whistleblowers from the major alphabet agencies, as well as big tech," Mikki explained. "Several of them were saying, 'Get ready. Something's coming. Any day now we're going to have another 9/11-sized event.'" A few weeks later, the pandemic hit.

CHAPTER TWO

PLANDEMIC 1

The supreme art of war is to subdue the enemy without fighting.

—Sun Tzu

The World Health Organization (WHO) announced the existence of a new coronavirus in China on January 9, 2020. By January 20, US airports were screening for the disease. The next day, the CDC confirmed the first US case. Day after day, the media announced new landmarks for the growing pandemic, but in Ojai—as in so many other parts of the country—it initially seemed like something that would be contained. One side of the media pronounced it the end of the world; the other side said that it was about to disappear. Neither was right, and nobody knew who to believe. *The Narrative* was being constructed in real time.

The first week in March, there were still no cases in Ojai's Ventura County, although seven residents had been tested (and tested negative). Despite that, by mid-March, Governor Gavin Newsom put a lockdown on the entire state of California. Mikki and the rest of the team continued their work from home and kept up communications on daily phone calls. As their research into the unfurling crisis continued, and questions arose, they sought out as many sources as they could to find raw, unadulterated information about the pandemic.

Before long, one vital source came to Mikki's mind: Judy Mikovits. Mikki and Judy had met almost eighteen months before the pandemic began. A mutual friend had introduced them as Judy began promoting a book she'd written about the medical industry and its failings.

"I liked Judy right away," Mikki said. "She's very raw. No show. No act. If she doesn't think something's funny, she doesn't laugh. She's just very real, with kind of an East Coast air about her that I respect."

Mikki considered doing a documentary with her at the time that they met, but he'd already committed to directing *The Narrative*. When his plans changed because of the pandemic, however, she was one of his first interviews.

***

"So, you made a discovery that conflicted with the agreed-upon narrative," Mikki said, staring intently at Dr. Mikovits as the cameras rolled.

"Correct," she smiled.

"And for that, they did everything in their power to destroy your life," he continued.

"Correct."

"You were arrested."

"Correct."

"You were put under a gag order. You were placed in jail. Yet, you sit here." Mikki continued. Dr. Mikovits nodded, a hint of sadness showing.

"Apparently their attempt to silence you has failed," he said. "I have to ask: How do you sit here with confidence to call out these great forces and not fear for your life as you leave this building?"

This interview—the first for the *PLANDEMIC* project—was not hyperbole. Depending on whom you speak with, Dr. Judy Mikovits is either remarkable and brave or disreputable and crazy. From the inner-most circles of science, medicine, and academia, she has become a symbol of conspiracy theories and bad science. Others see her as a crusader for truth. No matter how you interpret her path, it is clear that she paid dearly for daring to speak out.

Dr. Mikovits got her start working as a lab technician at the National Cancer Institute (NCI). (Much of the early research into HIV and AIDS was done there because one of the earliest-known symptoms of the disease—before they even knew what the disease *was*—was a rare form of skin cancer known as Kaposi's Sarcoma.)

In recent years, she has published books on important topics like vaccines, disease, and autism. In 2014, she published *Plague: One Scientist's Intrepid Search for the Truth about Human Retroviruses and Chronic Fatigue Syndrome (ME/CFS), Autism, and Other Diseases*, and in April 2020, she published *Plague of Corruption: Restoring Faith in the Promise of Science*. (Neither made much of a splash upon publication, although *Plague of Corruption* would skyrocket to #1 on Amazon and a place on the *New York Times* bestseller list after the release of *PLANDEMIC*.)

In short, when coronavirus struck the nation, Dr. Mikovits was just another scientist watching from the sidelines—albeit one with an unusual connection to the power players involved. In fact, given her history studying immunology, she could have been at the podium with Dr. Anthony Fauci and Dr. Deborah Birx, maybe even speaking in place of them. Why wasn't she?

That was just one of the questions that Mikki and his team hoped she could answer. Another question: Why, after decades working behind the scenes, was she suddenly speaking out against her former colleagues and the scientific and medical communities as a whole?

"Because if we don't stop this now," she told Mikki, "we can not only forget our Republic and our freedom, but we can forget humanity—because we'll be killed by this agenda."

It was heady stuff. According to Dr. Mikovits, it was a mass conspiracy spanning generations—and much of it could be traced back to one man: Dr. Anthony Fauci. The head of President Trump's coronavirus task force in 2020, Dr. Fauci also was the Director of the National Institute of Allergy & Infectious Diseases (NIAID) at the time that the AIDS epidemic was ravaging America.

As Dr. Mikovits told Mikki, Dr. Fauci's political angling at the height of the AIDS epidemic prevented French researchers from publishing their findings on HIV for well over a year after the original discovery—leading to unnecessary delays for a treatment, an accelerating spread of the virus during the peak of the epidemic, and millions of deaths. According to Dr. Mikovits, this is how it happened.

Working under researcher Dr. Frank Ruscetti, she said, "I was part of the team that isolated HIV from the saliva and blood of the patients from France." French scientist and 2008 Nobel Prize winner Luc Montagnier had already isolated the virus, which means they were able to separate it from biological samples like blood and grow it on a culture.

The American team was meant to simply confirm the French study. According to Dr. Mikovits, though, "Tony Fauci and Robert Gallo were working together to spin the story in a different way." Rather, it seemed they wanted to take credit for—and make profit from—the discovery.

At the time, Dr. Fauci also was running the NIH's program for HIV, which oversaw the NCI and Dr. Mikovits's team. Though she normally had little interaction with him, Dr. Mikovits had a run-in with Dr. Fauci as her team prepared to publish their findings on the virus.

Dr. Mikovits explained that one day, "Dr. Ruscetti was out of town and Tony Fauci says, 'We understand that you have a paper in press, and we want a copy of it.' And I said, 'Yes, there is a paper in press and it's confidential, so no, I will not give you a copy of it.' He started screaming at me. Then he said, 'Give us the paper right now or you'll be fired for insubordination.' And I just said, 'I'm sure when Dr. Ruscetti gets back, you can have the conversation.'"

When Dr. Ruscetti returned, however, Dr. Mikovits claims, he was "bullied into giving Fauci the paper." According to Dr. Mikovits, Dr. Fauci then held up the publication of that paper for several months, while Dr. Gallo "writes his own paper and takes all the credit." Dr. Mikovits was outraged. While the researchers played their political games, people were dying. "This delay of the confirmation literally led to spreading the virus around, killing millions," she said.

"It's still been crushing to me to think that I didn't know. . . . The virus didn't have to wait until '84 to be confirmed," she continued. "Think of how many people . . . The entire continent of Africa lost a generation as that virus spread because of the arrogance of a group of people."

One of those lost was Mikki's own brother, who died of AIDS in 1994. That same year, AIDS became the leading cause of death for all Americans aged twenty-five to forty-four. By 1995, more than 500,000 Americans would die from AIDS-related diseases. It was a certified epidemic, and one that the world had never seen coming.

The outbreak began quietly, with the mention of an "exotic new disease" in the *New York Native*, a gay newspaper in New York in 1981. By the next year, the CDC had formed a task force to track and confront the disease, which they coined AIDS (Acquired Immunodeficiency Syndrome). In 1983, two separate teams—one under American Dr. Robert Gallo and one under French researchers Françoise Barre-Soussi and Luc Montagnier—declared in *Science* magazine that they'd been able to isolate the novel retrovirus that was likely infecting AIDS patients.

The American team was led by Dr. Fauci, who had been at the NIH since 1968. In 1980, he was appointed the Chief of the Laboratory of Immunoregulation, and he became director of the NIAID in 1984. Dr. Fauci was responsible—and, at the time, notorious—for spearheading efforts to push through a new AIDS drug known as AZT (azidothymidine). In just twenty-five months, it was FDA-approved and being marketed as a miracle drug.

For Mikki, his brother's taking it was the decision that killed him. "He was killed by AZT, and there's no doubt about that at this point," Mikki told me. "I remember when this so-called miracle drug was released. My mom was so happy. She was very trusting in the leadership of the AIDS epidemic, particularly Anthony Fauci. At the time, there was no reason to doubt him."

"My brother believed he was getting a second chance at life," Mikki continued. "We expected his condition to improve, but as soon as he began taking AZT, he got much worse. We were told that his body was just acclimating and that he would get better in time." However, that time never came.

"He had migraine headaches," Mikki said. "He was constantly vomiting and too dizzy to stand. He was suffering so much that he began saying that he'd rather die. A couple of times he stopped taking the medication, and overnight he looked and felt better. It was confusing. His doctors warned him that even though he may feel better, if he went off the protocol for too long, he was going to die. Reluctantly, he stayed on the program.

"What do some of the worst-managed epidemics and pandemics, such as AIDS, Ebola, H5N1 (bird flu), and H1N1 (swine flu), have in common?" Mikki asked. "Dr. Anthony Fauci. How this man has maintained his position at the top of the medical pyramid is a mystery that deserves investigation."

Some will remember that Dr. Fauci was the one who set the world in panic when he predicted that "even in the best-case scenarios the bird flu (H5N1) will cause two to seven million deaths worldwide." To say that he was off is an understatement. The actual death toll landed in the hundreds. The damage caused by mass panic is impossible to quantify.

CBS investigative reporter Sharyl Attkisson was one of the few journalists brave enough to report the untold story. "We discovered through our FOI (Freedom of Information) efforts that before the CDC mysteriously stopped counting swine flu cases, they had learned that almost none of the cases they had counted as swine flu were, in fact, Swine Flu or any sort of flu at all!" she said. "In the end, no [CBS television news] broadcast wanted to touch it. We aired numerous stories pumping up the idea of an epidemic, but not the one that would shed original, new light on all the hype. It was fair, accurate, legally approved and a heck of a story. With the CDC keeping the true swine flu stats secret, it meant that many in the public took and gave their children an experimental vaccine that may not have been necessary."[1]

In March 2020, Children's Health Defense issued this reminder: "Fauci once shilled for the fast-tracked H1N1 influenza ('swine flu') vaccine, reassuring viewers in 2009 that serious adverse events were 'very, very, very rare.' Shortly thereafter, the vaccine went on to wreak havoc in multiple countries, increasing miscarriage risks in pregnant women in the US, provoking a spike in adolescent narcolepsy in Scandinavia, and causing febrile convulsions in one in every 110 vaccinated children in Australia."[2]

At the start of his decades-long career, Dr. Fauci was empowered to lead the HIV/AIDS epidemic. "He was given unearned credibility by being branded 'America's Top Doctor,'" Mikki insisted. "All the while, it was his suppressing of effective medicines and his pushing of deadly drugs that allowed AIDS to spread and kill. Drugs like AZT."

Mikki explained, "AZT is a dangerous and expensive drug that Dr. Fauci hailed as the miracle remedy while suppressing inexpensive, safe, and effective medicines from people in desperate need. He even backed AZT for pregnant women despite knowing there were serious risks to the fetus."

In November 1989, *Spin* magazine published an article by Celia Farber called "Sins of Omission." *Spin* founder Bob Guccione had this to say in a 2015 anniversary issue: "Celia unearthed hard evidence of the cold-bloodedness of the AIDS establishment pushing a drug that was worse than the disease, and killed faster than the natural progression of AIDS left untreated. AZT had been an abandoned cancer drug, discarded because of its fatal toxicity, resurrected in the cynical belief that AIDS patients were going to die anyway."[3]

According to former *Business Week* journalist Bruce Nussbaum, the man credited with creating AZT at the NCI under Fauci, Jerome Horwitz, was so disgusted with his own invention that he claimed it to be "so worthless that it wasn't worth patenting." With his name on the patent and a share of the profits, Fauci did not agree.[4]

Others were more direct with their accusations. In 1989, prominent AIDS activist Larry Kramer penned "An Open Letter to Dr. Anthony Fauci" in *The Village Voice.* In it he writes, "You are responsible for all government funded AIDS treatment research. In the name of right, you make decisions that cost the lives of others. I call that murder."[5]

Dr. Joseph Adolph Sonnabend, a highly regarded scientist and HIV/AIDS researcher, went on record to say, "I'm ashamed of my colleagues. . . . This is such shoddy science it's hard to believe nobody is protesting. The name of the game is to protect your grant. It's all about money. There are obviously financial and political forces driving this."[6]

Dr. Robert E. Willner, noted for his role in AIDS research and author of *Deadly Deception: The Proof That Sex and HIV Absolutely Do Not Cause AIDS,* joined a chorus of physicians who at that time were challenging the theory that AIDS was caused by HIV. In October 1994, during a press conference, Dr. Willner made the following statement as he injected himself with AIDS-infected blood on live TV: "I say to my friends, Fauci . . . and Gallo, and all the rest of those criminals, that this is for the sake of humanity and no other reason. . . . And this is in the hope that it'll save the lives of millions of individuals who will die because of the greatest lie ever told. . . . Indeed, it is the AIDS drug AZT that is the leading cause of AIDS today." Five months later, on April 15, 1995, Dr. Willner died of a heart attack.

Is Dr. Fauci following the same playbook today? Frontline doctors and infectious disease experts around the world are now speaking out against Dr. Fauci for suppressing proven medicines such as Ivermectin, Azithromycin, and Hydroxychloroquine.

When President Trump spoke favorably about hydroxychloroquine (HCQ) as a prophylactic against COVID-19, the corrupt factions of Western medicine went into panic mode and began turning the truth inside out. Why would the WHO, the FDA, Dr. Fauci, Dr. Birx, and countless other top doctors suddenly declare a 70-year-old, tried-and-true, safe and effective medicine to be "anecdotal" and "deadly"? It's baffling. That is, until you know the rules of that game.

The entire "plandemic" is propped up by what is called Emergency Use Authorization (EUA). Under the authority of EUA, existing laws can be overridden in the interest of public safety, laws that even after the emergency is over will remain. The moment a solution to the emergency arises, the EUA status is revoked and the pandemic is over. Those who are profiting from the pandemic will do most anything to keep it alive, even if that means allowing people to die. A harsh reality to embrace, I know.

As a side note, on July 28, 2021, *Axios* reported that "Pfizer expects revenue from the COVID-19 vaccine, co-developed by BioNTech, will reach $33.5 billion this year—a 29 percent jump from the previously estimated $26 billion."[7]

Convincing the general public that good medicines were bad didn't take much effort, as we've all been well trained to "follow the science." But how do you convince a generation of doctors and scientists that a medicine they've used with great success for decades has suddenly stop working?

First, you conduct bogus studies.

Second, you get top science journals to validate the studies.

Third, you get the media to echo the lie until it appears to be true.

ABC *World News Tonight*: "A new *Lancet* study shows patients hospitalized for COVID-19 face a higher risk of death if they take hydroxychloroquine. President Trump revealed this week he was taking the drug to prevent an infection."[8]

CBS News: "*The Lancet* medical journal has just published a study suggesting that treatment with a well-known antimalarial drug offers no benefit for patients with COVID-19. The study looked at chloroquine or its analogue hydroxychloroquine, which US President Donald Trump has been taking."[9]

MSNBC:[10] "Hydroxychloroquine, the antimalarial drug touted by President Trump, is linked to increased risk of death in coronavirus patients, according to analysis of 96,000 patients published in *The Lancet*."[11]

The media blitz created widespread panic, forcing clinical trials on hydroxychloroquine to be shut down before the drug could be officially proven effective for COVID-19.

According to an article in *The Scientist*, in October of 2020, "The study was a medical and political bombshell. News outlets analyzed the implications for what they referred to as the 'drug touted by Trump.' Within days, public health bodies including the World Health Organization (WHO) and the UK Medicines and Healthcare products Regulatory Agency (MHRA) instructed organizers of clinical trials of hydroxychloroquine as a COVID-19 treatment or prophylaxis to suspend recruitment."[12]

Numerous legitimate, independent scientists around the globe began to scrutinize the report in detail. That scrutiny led to serious questions about the integrity of the study, the authenticity of the data, and the validity of the methods the authors used. As it turned out, the lead coauthors of the study both have significant financial conflicts.

Dr. Mandeep Mehra, a lead coauthor, is a director at Brigham & Women's Hospital. Dr. Mehra and *The Lancet* failed to disclose that Brigham Hospital has a partnership with the biopharmaceutical company Gilead Sciences, which at the time was conducting two trials on Remdesivir, the prime competitor of hydroxychloroquine.

The database used to discredit hydroxychloroquine belongs to Surgisphere Corporation, whose founder and CEO is Dr. Sapan Desai. Dr. Desai, a lead coauthor of the study, flatly refused to disclose the data for independent confirmatory review. Furthermore, he refused to identify the participating hospitals, or even the countries.

In June of 2020, *Science* magazine reported, "Two elite medical journals retract coronavirus papers over data integrity questions . . . it focused on the safety and effectiveness of the malaria drug hydroxychloroquine for COVID-19, which had already become a political and scientific controversy, in large part because of Trump's embrace of the drug."[13]

In October of 2020, *The Scientist* reported, "At the heart of the deception was a paper published in *The Lancet,* on May 22, that suggested hydroxychloroquine, an antimalarial drug promoted by US President Donald Trump and others as a therapy for COVID-19, was associated with an increased risk of death in patients hospitalized with the disease. . . . The provenance of Surgisphere's database—if it even exists, which many clinicians, journal editors, and researchers have questioned—has yet to become clear. Most of Desai's coauthors admitted to having only seen summary data, and independent auditors tasked with verifying the database's validity were never granted access."[14]

In June of 2020, *Alliance for Human Research Protection* asked, "How did these studies, that were apparently designed to falsify the effects of a widely used drug, pass peer review in the world's premier medical science journals – *The Lancet* as well as *The New England Journal of Medicine?*"[15]

During a Texas Senate testimony on March 10, 2020, cardiologist and Professor of Medicine Peter McCullough testified that "85 percent of COVID patients given multi-drug treatment plan recover from the disease with complete immunity." McCullough added, "The pandemic could have been over by now if those who tested positive for COVID had been immediately treated before they fell ill enough to be hospitalized. He also says that thousands could have been, and still could be, saved if the treatment protocol he and other physicians use were not suppressed."

On August 24, 2020, during an interview with Mark Levin, of *Life, Liberty & Levin*, Dr. Harvey Risch, professor of epidemiology in the Department of Epidemiology and Public Health at the Yale School of Public Health and Yale School of Medicine, was passionate when he claimed, "This has gone on before. Now we have Dr. Fauci denying that any evidence exists of benefit [of HCQ]. The FDA has relied on Dr. Fauci and his NIH advisory groups to make the statements saying that there is no benefit in using Hydroxychloroquine in outpatients. And this is counter to the facts of the case. The evidence is overwhelming. Dr. Fauci and the FDA are doing the same thing that was done in 1987 and that's led to the deaths of hundreds of thousands of Americans that could have been saved by the usage of this drug! It's outrageous!"[16]

Dr. Fauci's dereliction of duty doesn't end with prescribing bad medicines. His go-to tool for testing for infectious disease is known as Polymerase Chain Reaction (PCR). PCR is currently being used universally as the gold standard test for COVID-19.

Dr. Kary Mullis, who won the Nobel Prize in Chemistry for inventing PCR, stated publicly numerous times that his invention should never be used for the diagnosis of infectious diseases. In July of 1997, during an event called *Corporate Greed and AIDS* in Santa Monica CA, Dr. Mullis explained on video, "With PCR you can find almost anything in anybody. It starts making you believe in the sort of Buddhist notion that everything is contained in everything else, right? I mean, because if you can model amplify one single molecule up to something that you can really measure, which PCR can do, then there's just very few molecules that you don't have at least one single one of them in your body. Okay? So that could be thought of as a misuse of it, just to claim that it's meaningful."

Mikki explained, "The major issue with PCR is that it's easily manipulated. It functions through a cyclical process whereby each revolution amplifies magnification. On a molecular level, most of us already have trace amounts of genetic fragments similar to coronavirus within us. By simply over-cycling the process, a negative result can be flipped to a positive. Governing bodies such as the CDC and the WHO can control the number of cases by simply advising the medical industry to increase or decrease the cycle threshold (CT)."

In August of 2020, the *New York Times* reported that "a CT beyond 34 revolutions very rarely detect live virus, but most often, dead nucleotides that are not even contagious. In compliance with guidance from the CDC and the WHO, many top US labs have been conducting tests at cycle thresholds of 40 or more. NYT examined data from Massachusetts, New York, and Nevada and determined that up to 90 percent of the individuals who tested positive carried barely any virus."[17] 90 percent!

In May of 2021, CDC changed the PCR cycle threshold from 40 to 28 or lower for those who have been vaccinated. This one adjustment of the numbers allowed the vaccine pushers to praise the vaccines as a big success.

In April of 2020, during an interview for *Uncover DC* with journalist Celia Farber, Canadian researcher, biologist, and president of Rethinking AIDS David Crowe said, "So, if you cut [PCR testing] off at 20 [cycles], everybody would be negative. If you cut off at 50, you might have everybody positive."

In May of 1996, during an interview with talk radio host Gary Null, Dr. Mullis revealed that his invention had been abused to falsify AIDS cases: "The number of cases reported went up exponentially because the number of tests that was done went up exponentially."[18]

Dr. Mullis went on to say, "This whole thing is a big sham. Guys like Fauci get up there and start talking. He doesn't know anything really about anything. And I'd say that to his face. Nothing. . . . He should not be in a position like he's in. Most of those guys up there on the top are just total administrative people, and they don't know anything about what's going on at the bottom. . . . Those guys have got an agenda, which is what we would like for them not to have, being that we pay for them to take care of our health in some way. . . . They make up their own rules as they go. They change them when they want to. Tony Fauci does not mind going on television in front of the people who pay his salary and lie directly into the camera."

Dr. Mullis believed that Dr. Fauci and others at the highest levels were all in on the sham: "They don't want people like me walking up and asking them those kinds of questions. And they're willing to go to great lengths to prevent that."

Kary Mullis died of pneumonia on August 7, 2019, seven months before the COVID-19 pandemic.

In August of 2021, a *Journal of Infection* report concluded, "In light of our findings that more than half [50–75 percent] of individuals with positive PCR test results are unlikely to have been infectious, RT-PCR test positivity should not be taken as an accurate measure of infectious SARS-CoV-2 incidence."[19]

During his *Uncover DC* interview with Celia Farber, David Crowe spoke out in defense of the late Dr. Mullis. "I'm sad that Kary isn't here to defend his work," he said. "He did not invent a test. He invented a very powerful manufacturing technique that is being abused." Just three months later, in July of 2020, David Crow died of cancer.

One year later, in July of 2021, the CDC quietly released this *Lab Alert* on their website: "After December 31, 2021, CDC will withdraw the request to the U.S. Food and Drug Administration (FDA) for Emergency Use Authorization (EUA) of the CDC 2019-Novel Coronavirus (2019-nCoV) Real-Time RT-PCR Diagnostic Panel, the assay first introduced in February 2020 for detection of SARS-CoV-2 only. CDC is providing this advance notice for clinical laboratories to have adequate time to select and implement one of the many FDA-authorized alternatives. In preparation for this change, CDC recommends clinical laboratories and testing sites that have been using the CDC 2019-nCoV RT-PCR assay select and begin their transition to another FDA-authorized COVID-19 test."

Days after the CDC made that shocking announcement, *The Times* (UK) reported that "George Soros and Bill Gates are part of a consortium acquiring a British developer of rapid-testing technology, including for Covid-19 and tropical diseases, to turn it into a social enterprise."[20]

In *PLANDEMIC*, Dr. Mikovits reveals the policy she believes is at the core of this medical corruption. When asked about the troubling subject of "conflicts of interest," she offered a clear and concise solution: "Repeal the Bayh-Dole Act."

In 1980, the Bayh-Dole Act (also known as the Patent & Trademarks Law Amendment Act) was passed by Congress, granting scientists at federal agencies and universities the right to claim personal ownership of inventions or discoveries that were made with federal funding. In short, taxpayers paid millions to fund these discoveries, which the scientists could then sell to the same pharmaceutical companies that would charge the taxpayers for their medicines.

The result should have been predictable: Today, universities obtain sixteen times as many patents as they did in 1980. In many cases, critics say, the drive for dollars can push scientists toward work that will make them rich—instead of work that will help humanity.

"That act gave government workers the right to patent their discoveries, to claim intellectual property for discoveries that the taxpayer paid for," Dr. Mikovits explained to Mikki. "Ever since that [development] in the early eighties, it destroyed science and allowed for the development of major conflicts of interests."

Dr. Mikovits saw that firsthand in May 1985, when a patent was awarded on Dr. Gallo's work around HIV. (Remember, Dr. Fauci had delayed the publication of Dr. Ruscetti and Dr. Mikovits's study so Dr. Gallo could publish his first.) For his part, Dr. Fauci and future CDC head Redfield "were working together to take credit and make money," Judy claimed. The duo had patents for a therapy known as IL-2 therapy. Therefore, Judy alleged, they "tailored" studies to support that patented treatment, although it "was absolutely the wrong therapy."

According to Dr. Mikovits, if there had never been a Bayh-Dole Act, and scientists like Dr. Fauci didn't have a financial conflict of interest, a better treatment could have emerged sooner. She said, "millions wouldn't have died from HIV." It was a race for profit instead of a race to the cure. (Perhaps not coincidentally, Dr. Fauci is now very outspoken about his support for issuing patents for the COVID-19 vaccines, although not doing so could help millions in underprivileged nations.)

The American public was largely unaware of how scientists were cashing in until a 2005 investigation by the Associated Press found that US National Institutes of Health researchers like Gallo had received nearly $9 million in royalties on their taxpayer-funded patents. In particular, they noted that Fauci and his deputy, Clifford Lane, received payments that were related to their discoveries in the treatment of HIV and AIDS back in the 1980s. Dr. Fauci claimed he had donated all proceeds to charity, although he never provided any public proof of that.[21]

Although this report was the first time that many heard about the conflicts of interest that were rampant at the NIH, government officials were aware of how it could unduly influence the focus and nature of research. In 2000, the head of the Department of Health and Human Services, Donna Shalala, instituted a new policy requiring scientists to disclose financial interests related to their work. However, nothing tangible was done to ensure compliance with that policy—at least, not until the Associated Press shined a spotlight on the issue.

Another consequence of Bayh-Dole and the lure of patent royalties was the undue influence of the wealthy in the world of American science. People like convicted sex offender Jeffrey Epstein gave millions to universities and other organizations for research—organizations that desperately courted their largesse. Over time, these individuals were given a kind of academic credence, which in most cases they did not deserve.

In the case of Bill Gates, for example, Dr. Mikovits said, "Nobody elected him. He has no medical background. He has no expertise. But we let people like that have a voice in this country while we destroy the lives of millions of people." Indeed, Gates has reinvented himself in the 21st century. Once known primarily as a tech entrepreneur and the creator of Microsoft, he has focused on philanthropy since founding the Bill & Melinda Gates Foundation with his wife in 2000. The largest foundation in the world, holding $51 billion in assets, it cites "enhancing healthcare" as one of its primary goals. To Bill and Melinda, that has primarily meant vaccines.

From 2009 to 2015, for example, the organization receiving the largest amount of funding from the Foundation—to the tune of more than four billion dollars—was GAVI, the Vaccine Alliance. GAVI was founded in 2000, the same year as the Gates Foundation. It also was founded by the same people, Bill and Melinda Gates, with an opening donation of $750,000,000. On its website, GAVI brags about having vaccinated "822 million children in the world's poorest countries, preventing more than 14 million future deaths."[22]

In addition to existing diseases, GAVI and Gates were focused on developing vaccines for illnesses that weren't even rampant yet. For example, in 2017, the organization claimed that it had organized "the largest coalition to prevent a pandemic," the Coalition for Epidemic Preparedness Innovations (CEPI). CEPI was funded—of course—by the Bill & Melinda Gates Foundation, and also by the German, Norwegian, and Japanese governments.

Also on board for GAVI and the Gateses' pandemic vaccine project was the Wellcome Trust. Founded by a British pharmaceutical magnate, Wellcome funds scientific and medical research around the world. As a result, they have more than a dozen patents in the United States, including one for a sprayable rotavirus vaccine that was issued in February 2020, mid-COVID. Two months after that patent was issued, a researcher at Indiana University Bloomington was already suggesting that rotavirus vaccines could perhaps be reengineered to prevent COVID-19, particularly if administered to children. Coincidence? Probably not, as we'll see.

Vaccines can save lives, but they're also big business. It's important to be aware of that distinction, and believing one doesn't mean you can't believe the other. For example, when Mikki asked Judy if she was "anti-vaccine," she was adamant: "Oh, absolutely not. In fact, vaccines are immunotherapy, just like interferon alpha is immunotherapy. So, I'm not antivaccine. My job is to develop immune therapies. That's what vaccines are, at least, when they're made safely."

The issue at hand, however, the looming reality that evades normal people of the world, is that the very people advocating for vaccines stand to make millions of dollars from their implementation, due to patent ownership. No matter how ethical they may be, their advice is therefore compromised. When you stand to make money from a medical treatment that you've developed, it can be nearly impossible to provide unbiased guidance on the efficacy of that treatment, even if you're well intentioned. In any other industry, this is Business Ethics 101. For some reason, however, within the medical industry, these incestuous structures are obvious, unethical, and largely ignored.

What's worse, anyone who dares to question these conflicts of interest is censored, attacked, discredited, and pushed to the back of the press room—if not evicted altogether. Many of our doctors and scientists have somehow made themselves beyond reproach and even beyond questioning, when the very nature of their work requires relentless examination and query. How did we ever get here?

Here's a perfect example of how these forces currently interact. As COVID-19 raged in April 2020, Bill Gates made the rounds of the press circuit, talking to anyone who'd have him. Despite the fact that he has no medical education, he was confident in his prescription for the public: "Normalcy only returns when we largely vaccinated the entire global population," he said.[23] At the same time, the Bill & Melinda Gates Foundation had already donated millions of dollars towards COVID-19 vaccine research: $3.6 million to SK Bioscience in South Korea, $1 million for Shanghai Zerun Biotechnology in China, and more than $4 million for Biological E. Limited in India, for example. The results of these companies' research have not yet been publicized. If and when they pursue a patent, however, the Bill & Melinda Gates Foundation could cash in.

Although traditional patent ownership goes to the inventor, it also may be transferred to what is called the "assignee," or the entity that has the property right to the patent and that therefore will receive royalties from its use. The Bill & Melinda Gates Foundation already is listed as the assignee on several US patents that have been generated from the scientific research they've funded. So, if Bill Gates isn't recommending vaccines as the cure due to his medical knowledge—because he doesn't have any—then what is his motivation?

During an interview on CNBC, journalist Becky Quick asked Mr. Gates, "You've invested 10 billion dollars in vaccinations over the last two decades, and you figured out the return on investment for that. It kind of stunned me. Can you walk us through the math?" Bill Gates replied, "Over a 20 to 1 return. So if you just look at the economic benefits, that's a pretty strong number compared to anything else."[24]

Bill Gates isn't the only person with financial ties to the very treatments he has attempted to popularize—far from it. Doctors and scientists throughout the medical-industrial complex have thousands of patents in their names, and profits flowing to them regularly.

Dr. Mikovits's interview—and really, her entire career—encapsulates the story of how money, science, politics, the media, and power intersect in the United States. In the midst of a pandemic, there was no more important story to be told.

"In the very beginning, I thought, *If anything, I'll edit this together for Judy. I'll pay for it out of my own pocket, and I'll give it to her as a gift for all she's given,*" Mikki said. "So, at least if she's going to go out and try to get her movie made, or try to get the word out there about what she's been witness to, she can use this interview to further her cause. And because it was apparent at the time that Anthony Fauci was going to resume his post as 'America's doctor,' I also started to feel the need to at least give the people the hidden information so they could make informed decisions about their health and future." That small favor soon grew into something much bigger.

"At the time, we were in a trend where the #MeToo movement and 'Believe Women' was very prevalent, which led me to think, 'Well, if that's a real mission statement that we're truly going to live by, then I think it's only fair that people have a chance to hear *this* woman,'" Mikki said. "It's important to give voice to people who have been wronged, especially women within the boys club of science."

When the clip was finished, Mikki was certain: The world had to see it.

CHAPTER THREE

Debunking the Debunkers

All truth passes through three stages. First, it is ridiculed.
Second, it is violently opposed. Third, it is accepted as being self-evident.

—Arthur Schopenhauer

Mikki and his key researcher, Nathaniel, worked fast and furious on *PLANDEMIC 1*. "Before hitting send, we carefully researched each claim made by Dr. Mikovits. We were confident that the information she provided was accurate, or at the very least unsettled science," according to Mikki.

On May 4, it was go time. "Before we hit send, we shouted, 'May the 4th be with us!' It was our playful way of asking for help and guidance from the Universe." Holding their breath in the Ojai office, Mikki and

Nathaniel watched as the 26-minute video was uploaded to Facebook and YouTube.

"The video went viral beyond our wildest expectations," said Mikki. "I expected that it might get a few hundred thousand views or maybe even a million or so. With the amount of people who have been wronged by bad medicine, I knew that people would care enough to share it. I hoped that people were intuitive and sensitive enough that they could feel Judy and know that there was truth in her words."

The video swiftly generated hundreds of thousands, then millions of views. Within a week, it reached over 100 million views as people around the world promoted it on their own social media platforms. Mikki and the team knew that the video would make a splash, at least in some quarters, but this was a tsunami.

Little did anyone know, it was just a small team tucked into a non-descript office on top of a mountain who had dropped this bomb-shell upon the world. Around them, the entire town of Ojai was continuing on with a normal day. Probably even the crew down-stairs at the coffee shop had no idea what had been unleashed above them. All they had ever really wanted to do was to share a voice and provide an alternative viewpoint to one of the most consequential moments in, not only American, but in global history.

"The initial response was overwhelmingly positive. People from all over the world were sharing the video and leaving comments of gratitude for Dr. Mikovits and her courage," said Mikki. That is, until the critics descended. US media and fact-checkers took unprecedented measures to smear and destroy the messengers.

"They are alchemists stuck in reverse, turning everything beau-tiful into something ugly" Mikki said. "Overnight, every good deed I'd ever done was twisted into evidence that I was not to be trusted. Somehow, online videos of my 50th birthday party was branded as a cult gathering. My wife, one of the kindest women alive, was smeared as a 'Jezebel.' Triggered by those unfounded judgments, concerned citizens demanded that our children be taken into protective custody."

Mikki continued, "Watching that unfold reminded me of a media assignment I was part of in 2007. We filmed a series of innocent moments happening at the Santa Monica Pier: A mother breastfeeding. A clown making balloon animals. Lovers kissing. Kids playing arcade games, etc. From this footage we created two versions of the same short film. Version One was scored with a fun soundtrack, while Version Two was scored with horror film tracks. Despite the fact that the imagery was identical, when polled, the audience saw two totally different movies. The breastfeeding mother went from 'beautiful' to 'creepy.' The 'cute' clown became 'scary.' The kissing lovers, 'disturbing.' The kids in the arcade were 'in danger.' A soundtrack. That's how easy it is to manipulate an audience. The masters of propaganda know this better than anyone."

Mikki added, "As a veteran of media production, I wasn't surprised by the dirty tactics they were using to smear our good names. What did shock me was to see so many citizens so easily fooled by these tactics. People who were fierce supporters just days before were suddenly posting public apologies for having shared *PLANDEMIC*.

"The same people who just hours prior were flooding my inbox with virtual high-fives began to publicly distance themselves from me. I reached out to a few of these people to ask, 'Do you honestly trust the media over someone you've known for twenty plus years?' Like robots, they responded with programmed talking points: 'Your movie is dangerous. It's going to kill people. Bill Gates and Anthony Fauci are heroes. A vaccine is our only hope.'"

He continued, "The media and those who control it have done serious damage to our collective psyche. That said, if the number of apologies I'm currently receiving on a daily basis is any indication of a turning tide, The Great Reset is being replaced by The Great Awakening."

I was one of the people who turned up my nose at *PLANDEMIC*, scrolling by it on my Facebook feed without stopping to watch. The people who were sharing it and the reception by others I trusted were enough for me to avoid it, I thought. I was the kind of person who rolled my eyes at "antivaxxers" and nodded approvingly at "Believe Science" Tweets. Years of reporting had taught me not to trust the government, but I also knew that every major news story brought out its share of crazy conspiracy theorists. I thought that the *PLANDEMIC* people were more of those misguided kooks.

Then, I was assigned a feature on sanitizing products. Keen to understand the science behind the virus and how to stop it, I went to the original source documents—scientific research studies. What I read there directly challenged much of the guidance being offered by the CDC and WHO (guidance that, if you were paying attention, kept changing). Among the takeaways? Masks can be harmful. Washing your hands has limited usefulness. A safe and successful vaccine for the virus would not be created anytime soon. I watched as the CDC, the WHO, and other health organizations changed the guidance on their websites in real time, with no mention of the fact that different information had been there the day before. It was clear that there was a different narrative unfolding from the one being presented in the media.

Not long after that, I decided to watch *PLANDEMIC* for myself. I went down the rabbit hole of the information contained within. The result—as you'll see—is this book. Not many of my fellow journalists were willing to do the same work.

"I've had probably a couple dozen people, including some journalists, who have actually taken the time to do their own research," Mikki told me. "One hundred percent of the people who have done that have come full circle to say, 'What the hell is going on here? How can this film be completely 'debunked,' yet everything checks out?'"

***

By branding the movie as "debunked," the controllers of the global narrative coerced citizens to look away without further inquiry. For Mikki and his team, it seemed like every other minute a new alert would drop into their inboxes with a scathing critique by yet another news outlet. Many of them missed key points of the argument in their haste to tear it down, but the *PLANDEMIC* team tried to maintain a spirit of open-minded collaboration, nonetheless, writing to them to explain and request that the critics update their own pieces.

It seemed like no one had any curiosity or willingness to challenge the party line on the pandemic. They were met with a brick wall of intractability, and it was about to get worse. Of course, the film was intended to be provocative. They wanted to break through the media noise and get through to people. That didn't mean that it was false, however. The truth can often be one of the most provocative topics of all.

On May 6, two days after the video was released, the team was hit with another atom bomb: their video was vanishing. On Facebook, YouTube, Twitter, and all around the web, supporters were reporting that their videos were being taken down. Something strange was afoot.

A Facebook representative would later tell the *Los Angeles Times*, "Suggesting that wearing a mask can make you sick could lead to imminent harm, so we've removed the video."[1] That was in reference to a claim Dr. Mikovits made in the video, that mask wearing "activates the virus," and is more dangerous than going without a face covering. They seemed to have forgotten that the CDC, Department of Health and Human Services, Surgeon General, and other governmental leaders had told everyone *not* to wear masks at the beginning of the pandemic.

Whatever the logic, other sites were quick to follow, and as quickly as the video had caught fire, it was snuffed out. YouTube justified removing the video by saying they regularly remove "content that includes medically unsubstantiated diagnostic advice for COVID-19." Mikki and the team weren't sure what about the video was "diagnostic," but the video platform wasn't quick to illuminate its decision.

At Vimeo, they claimed that they were standing "firm in keeping our platform safe from content that spreads harmful and misleading health information. The video in question has been removed by our Trust & Safety team for violating these very policies."

Twitter seemed like it would be the only platform that would allow *PLANDEMIC* to live on, and the video continued to hold on there even after it was struck down from other sites. At the same time, however, hashtags for *#PLANDEMICMovie* and Dr. Mikovits's book, *#PlagueofCorruption,* were trending across the globe before suddenly disappearing from searches and trends. On Google, it was all but invisible.

Still, the video had made its mark. The *New York Times* wrote an article describing the impact of the video online, that it had dwarfed other trending topics, such as "*The Office* reunion," the release of Taylor Swift's new video, and the Pentagon's major announcement confirming the existence of "aerial phenomena."[2]

"Thanks to the power of the people and the censors for stirring curiosity, *PLANDEMIC 1* reached over one billion collective views, setting a world record," said Mikki. The digital censorship was another twist in the story that they didn't see coming, even though it was something Dr. Mikovits had warned them about. Like so many people, they thought, "It won't happen to us. We're telling the truth." Although nothing in the video could be proven false, it was shut down simply for questioning the dominant narrative.

In an effort to validate and clarify the most contentious claims made by Dr. Mikovits, Mikki and his team began to work on a follow-up interview. One of the major critiques was directed at Judy's implication that wearing a mask "reactivates" the virus. This one point had many critics dismiss the rest of what the video had to say. Facebook removed the video because of this claim in particular. In reality, though, Facebook and the rest of the critics spoke too soon.

Though Judy's language may have been imprecise, the fact is, wearing a mask can indeed make sick or recovering people sicker. An April 2020 article in the *New England Complex Systems Institute* journal by MIT scientist Yaneer Bar-Yam explains:

> A strategy to reduce COVID transmission is to wear a mask. However, for a person who is sick, wearing a standard mask can lead to increased rebreathing of viral particles. . . . Rebreathing coronavirus particles by an infected individual who exhales them may be harmful in accelerating COVID-19 progression. Infections initially occur due to inhaling coronavirus particles or touching the face. Once the infection is established in the nose or lungs, the virus replicates and particles are sneezed, coughed, or breathed out. These particles are capable of infecting others, and can also be rebreathed. Disease progression is a competition between viral replication and elimination by the immune system. Rebreathing increases the amount of virus (viral load), and can add new locations of infection in the lung.[3]

In patients with COVID-19, viral shedding can continue for up to thirty-one days, according to other studies, even as the patient feels they have recovered. If those viral particles are rebreathed because of mask use, the individual may get an entirely new lung infection.

In addition, "since 80 percent of cases are mild," Yar-Bam continued, reduced rebreathing could cut down on the number of people "who progress to severe cases and the overall impact of this disease." He suggested weighing the risk of rebreathing in any situation against the risk of breathing viral particles onto people around you. However, that would require the use of common sense, which the government does not appear to trust most people to utilize.

So, was *PLANDEMIC* really arguing, like the naysayers claimed, that masks *cause* the virus? Not quite. The truth was far less provocative. Ultimately, Dr. Mikovits clearly states that wearing a mask can *reactivate* the virus in the manner described above. It was a straightforward comment that was spun into something more outrageous. That could be said for so many other points in *PLANDEMIC 1*.

CHAPTER FOUR

PLANDEMIC 2

They who can give up essential liberty to obtain a little temporary safety deserve neither liberty nor safety.

—Benjamin Franklin

July 2020
Ojai, CA

Never before has there been so much transparency and openness to new ideas. Never before has there been such tribalism, misinformation, and virulence in the name of truth. The Internet has made both possible, and perhaps no one was more aware of that Catch-22 than the *PLANDEMIC* team were in the summer of 2020.

As the coronavirus continued to rage among us, and some of the most chilling predictions of the film came true, they watched the trolls and critics continue to slander and smear them. Most often, it was for things they'd never actually said.

"The media made a huge fuss over the fact that we used stock footage to visually illustrate a couple of moments within Dr. Mikovits's story," Mikki explained. "As Judy told the story of her arrest, lacking the actual footage, we used a clip obtained from a stock library. Anyone with a little knowledge of basic documentary filmmaking knows that the use of what's called 'B-roll' is a common practice. The proper question is 'Did we portray the re-creation honestly?'

"According to Judy, the clip we used was mild in comparison to the actual police raid that took place at her home. So, if anything, we down-played the moment. Still, critics insisted that this was proof that we had fabricated the entire story."

Overall, it seemed, most of the criticisms had nothing to do with their argument: They chose to focus on the tearing down of Dr. Mikovits's character. Yes, she's complicated. But that does not make her a liar. The bottom line for many people, it seemed, was "But, she's been *arrested*." People didn't care that the *PLANDEMIC* team *knew* all about that going in. In fact, they had the arrest warrant. A warrant that was never signed, making it invalid. Just like she claimed, Dr. Mikovits never had a single charge filed against her. None of that mattered. Judy Mikovits had already been ruled guilty in the court of public opinion.

"Like all of us, Judy Mikovits is imperfect," Mikki said. "Because of what she's been through, her emotions can at times override her ability to communicate like one might expect a scientist at her level to communicate. To magnify these human flaws, however, is to distract from the content of her testimony—a grave error."

To avoid stepping in that trap again, Mikki decided to feature other whistleblowers in *PLANDEMIC 2.* "Knowing we were under the microscope of the world, this time I hired a team of researchers to ensure that every claim made in Part 2 was unassailable. I also brought on my friend and producer Erik, and my long-time creative partner Gabriel," Mikki explained. "Under the lockdowns, we began conducting interviews over Zoom."

"After the way the first *PLANDEMIC* was attacked, I was surprised that so many high-level professionals were willing to risk their careers and lives to work with us. It was an honor to interface with top virologists, immunologists, infectious disease experts, and even a couple of renowned Nobel Laureates. I saw it as a testament to the severity of corruption within the medical industry that such accomplished professionals were willing to take such a risk. I learned so much through that process!" Mikki admitted.

"I had completed well over thirty interviews when I began receiving messages from friends suggesting that I look into a patent expert named Dr. David Martin. One of my researchers, Sean, was insistent that I interview Dr. Martin. He shared one of David's videos with me, and to be honest the information went over my head. I couldn't fully comprehend the impact of the information Dr. Martin was sharing. Trusting my team, I agreed to a Zoom call."

That call started off like every other interview Mikki had done. "Halfway into the interview, as David was on a roll, I leaned over to look around my computer monitor to make eye contact with Erik," Mikki shared. "I mouthed the words 'Let's stop.' Erik whispered, 'Why, what's wrong?' I said, 'This guy is brilliant. We have to do this one in person. Get him on the next flight.'"

Dr. Martin was in Ojai the next day. The information he had shared was so precise and critical that it changed the entire narrative of the project. I later asked Dr. Martin what was going through his mind when he agreed to go all in on the project. At a time when everything *PLANDEMIC*-adjacent was considered beyond controversial, what motivated him—as a financial analyst and researcher—to associate himself so publicly with them? "I figured that they had heard a couple of my YouTube videos and come away with the same conclusion as many viewers: This is either absolutely crazy, or some of the most important information that we can get across. It's likely that both options were somewhere in the consideration," he said.

Dr. Martin explained, "As I was flying out to California, what was going through my mind was that I'm not 'The COVID guy.' I'm a guy who talks on CNBC and Bloomberg about the markets. I'm a guy who testifies in Congress about criminal conspiracies, tax fraud, abuses of public trust, and all kinds of other things."

He continued, "Whether the storyline is the abuse of breast cancer genes, white collar crime involving corporations and universities colluding to defraud the government, intelligence failures justifying the war in Iraq, or COVID-19, I have the same story, no matter who I›m talking to. So that dispassionate part of me was just thinking, *I'm going to do another interview where I recite the facts of another time when patent information happened to be the thing that broke open a big story that matters to humanity.*"

By the time Dr. Martin arrived in Ojai, the expanded film crew was set up and ready for the interview that would become the throughline of a much longer and more detailed movie they would call *PLANDEMIC: INDOCTORNATION.*

Mikki invited Dr. Martin to sit in the stool across from him. After some small talk and technical checks, it was time. Dr. Martin began. "I'm the developer of linguistic genomics," he explained, "which was the first platform on which you could determine the intent of communication rather than the literal artifact of communication."

A niche field in the development of artificial intelligence and communications, linguistic genomics is designed to uncover the intent of communication versus the literal translation. You often see this at work in search engine results, for example, as the algorithms have become much more intent-driven over time.

For example, if you search "restaurants," your first result will not be a definition, or a cultural history of restaurants. The search engine understands that your intent is most likely to find restaurants near you, and that's what it will serve up. If you search for "Oscars," practically no search results will be related to an individual of that name. Instead, the search engine understands that you are likely searching for the Academy Awards and will deliver you results around that intent. The "literal artifacts of communication"—your words—are not the same as your intent.

On a higher level, computers can analyze huge portions of text communication to flag underlying messages and trends that humans can't even recognize or comprehend. Dr. Martin explained, "We've also used that technology for a number of other applications in defense, intelligence and finance," such as flagging tax fraud based on deviations and patterns outside of a normal human's scope of understanding. Dr. Martin's company also keeps a running algorithm intended to pinpoint situations where international actors could be making decisions that would put all of humanity at risk.

"We maintain a series of inquiries into every individual, every organization, and every company that is involved in anything that either blurs the line of biological and chemical weapons or crosses that line in any of 168 countries," he said.

Most relevant to Mikki was Dr. Martin's research on US patents. In 1998, Dr. Martin founded M-CAM, Mosaic Collateral Asset Management. "The goal of the company was singular. It was to provide a mechanism for banks to use intangible assets—patents and copyrights and trademarks and things like that—as collateral for lending," he explained. "We thought—naively, as it would turn out—that the patent office would be one of those relatively boring old institutions that wasn't really overly corrupt. We thought wrong."

In 1998, IBM signed a contract with the United States that facilitated the digitization of one million patents. For researchers like

Dr. Martin, this was a gold mine. Like never before, the history of innovation in America had been converted into a data set that could be analyzed from infinite perspectives. Researchers could identify trends that could catalyze new discoveries, or roadblocks to innovation that could be removed. What he found, though, was much more surprising—and chilling.

"Approximately one-third of all patents filed in the United States were functional forgeries, meaning that while they had linguistic variations, they actually covered the same subject matter. In other instances, when a company's patent was expiring or an inventor went from one firm to another, they would duplicate a filing in a process referred to as 'double patenting.'" That is, they would patent their project again, which is against the law. It was clear that businesses were engaging in all kinds of subterfuge under the staid cover of the US Patent Office.

Dr. Martin and his team actually read the substance of the patents, and not just the abstract—something few Americans had ever done. What they found only solidified their growing concern that scientists and big business were pulling one over on the American public.

"Within these documents, there are a lot of things being disclosed that are not what the title or the abstract or any superficial level of the patent actually was," he said. "You know, patents for a nuclear reactor that were examined in the patent office, by the patent examiners for bathroom fixtures. I was thinking, *This can't be real.* It was real."

In May 2001, Dr. Martin brought his findings to Congress for a hearing in the Subcommittee on Courts, the Internet and Intellectual Property of the Committee on the Judiciary of the House of Representatives. Then-Representative Howard Berman of California called Dr. Martin's findings "astonishing," as other members of the Committee, including Lindsey Graham, listened, rapt.

Also listening? The companies whose patents Dr. Martin was exploring. "These companies realized, 'Ah, oh, we may be in trouble,'" he explained. "And so, what they did was, they started donating their patents to universities and taking hundreds of millions of dollars of tax deductions as a result."

Here's how it would work. First, a company like DuPont or Monsanto would donate a patent to the university. The company would assign almost any value they wanted to the patent, perhaps saying that it was worth $50 million. They'd use that $50 million "donation" for their own tax benefit, and the university could then turn around and tell the government that they had received a matching grant of $50 million from a company. They could use that to then get $50 million in real cash from people like Dr. Fauci, since federal grant rules require universities to prove that they have industrial partnerships.

Dr. Martin continued, "Billions of dollars were being stolen and extorted from the government using the pen." No one had any idea.

"We wondered what other criminal enterprises and what other illicit activities were being hidden," he said. "The amazing thing is the audacity of criminal organizations. They hide their actions in plain sight. They hide them in places where nobody will ever think to look. Not surprisingly, patents—because nobody reads them—happen to be a good place to conceal illicit activities."

That included activities by the government. For example, Dr. Martin explained, "You have people talking about the fact that the United States is only focused on its biologic agent program for defensive purposes. Then, you see a United States patent for a blast-resistant pathogen fired from a rocket-propelled grenade. Last time I checked, a blast-resistant pathogen from a rocket-propelled grenade is hardly a vaccine in a syringe."

The words tell the story. Almost without trying, Dr. Martin had stumbled on the ultimate source document, a map showing exactly what the US government, big business, big science, and academia were *really* up to. Layers below all of the press releases, the patents were where the truth came to rest. Dr. Martin kept digging to find it.

"The only place you can go to validate the things that we actually uncovered are the digital fingerprints of the perpetrators of the crime— the filed patents themselves. If you get the words that people use, then you can use that to track down their grants and you can try to track down their affiliations," he said.

"Before long, you see that the Patent Office, the CDC, the FDA, the NIH, and the National Science Foundation are all in this massive collusive network, which is essentially a way to take public funds and underwrite corporate programs, and—probably most egregiously—pay exorbitant amounts of money to universities that rely on federal grants as one of their primary funding sources. Ultimately, the patent represents the commercial greed of an individual or organization, because what they're trying to do when a patent is filed is the obstruction of the free market, by definition. As a result of that, there is a high incentive to obstruct free markets across the system. And there is a high incentive to lie about it. And it turns out that when nobody was watching the store, both of those happened."

Before interviewing Dr. Martin, the *PLANDEMIC* research team conducted a thorough background check to see whether he was for real, or yet another conspiracy kook who had made the Internet his soapbox. Without exception, word came back: *David is the real deal.* What's more, when they told him that he'd checked out, he wasn't offended, as others might have been. Instead, he said that it was par for the course at this point.

"If you take your work seriously, you check your sources. That's pretty much the universal response," he said. "People find it so incredible that it's this egregious and this visible. They doubt that it could be possible."

At heart, Dr. Martin reminded me, his company deals with investments. "Because of SEC rules and banking rules and international rules, I have to maintain what's called an FBI-level chain of custody of all of our documents," he explained. "So when I say something exists, I don't have a hunch. I know what I'm saying is exactly what I'm saying. The fact is, that's not a standard most people live with."

"When people say to me, 'That's unbelievable,' I remind them that this isn't about belief," he continued. "It's about information that is—in most cases—hard for the average person to find. Even if you know that $191 billion in federal funds went out through the NIAID, and even if you know that a certain amount went to China, and a certain amount went to NGOs, how can you ever track what it was spent on? Well, patents are the best way to do it."

In the late 1990s, a new and puzzling trend was emerging in that sea of data. During 1999 alone, fifty-nine new patents were issued for medical discoveries related to the "coronavirus" disease family. Where was this new global interest coming from? More important, where would it lead? Above all, why, in 2002, did the University of North Carolina patent a recombinant version of coronavirus that had been adapted specifically to target human lung cells months before the invention of SARS?

Hong Kong, China
March 2003

The year 2003 was a dark one in Hong Kong. In February, the government proposed a bill that would bring back harsh antidissident regulations not seen since the era of British colonial rule. Intended to "prohibit any act of treason, secession, sedition, subversion against the Central People's Government," it was interpreted as a free pass for government officials to crush their detractors. Protests erupted, and the eyes of the world were watching. What came next, though, was even more devastating.

On March 11, Hong Kong recorded its first case of SARS-CoV1, a coronavirus causing respiratory infection. In the first three months alone, nearly 2,000 cases were identified, and hundreds died. SARS was entirely new and had never before been seen in humans or in animals. The international medical community swung into action. According to Dr. Martin, it wasn't just to save lives.

"In 2003, the Centers for Disease Control saw the possibility of a gold strike, and that was the coronavirus outbreaks that happened in Asia," he said. "They saw that a virus they knew could be easily manipulated was something that was very valuable, and, in 2003, they sought to patent it. They made sure that they controlled the proprietary rights to the disease, to the virus, and to its detection and all of the measurement of it."

In response to Dr. Martin's claims in *PLANDEMIC: INDOCTOR-NATION*, CDC spokesman Llelwyn Grant told the Associated Press (AP) that their intention in filing a patent for coronavirus in April 2003 was to prevent *other* bad actors from doing just that. "The whole purpose of the patent is to prevent folks from controlling the technology," he insisted. "This is being done to give the industry and other researchers reasonable access to the samples."[1]

While his argument may make sense at first, the reality is that it's patently false. Publishing the science would make it part of the public domain, and therefore *not* patentable—by anyone. Patents are not about protecting the science. They are solely focused on control and commercial gain. Plus, if the CDC were so intent on making the research information public, why would they file a request at the patent office to keep the patent application secret?

To Dr. Martin, that much was obvious. "We know that Anthony Fauci, that Ralph Baric, that the Centers for Disease Control, and the laundry list of people who wanted to take credit for inventing coronavirus were at the hub of this story," he said. "From 2003 and 2018, they controlled 100 percent of the cash flow that built the empire around the industrial complex of coronavirus."

What's more, the very patent itself was in a gray area, legally speaking. "Under 35 US Code, Section 101, nature is prohibited from being patented," David explained. "Either SARS coronavirus was manufactured, therefore making a patent on it legal, or it was natural, therefore making a patent on it illegal. If it was manufactured, it was a violation of biological and chemical weapons treaties and laws. If it was natural, filing a patent on it was illegal. In either outcome, both are illegal."

At the time the patent was filed, it was certainly controversial. Labs around the world scrambled to file their own competing patents in the hopes of getting credit for their early discoveries. One company working on a vaccine suggested that if they didn't get the patent, they'd stop their research.

"If we didn't have patent protection, we wouldn't invest in the research," CombiMatrix President and Chief Executive Amit Kumar said in one interview.[2]

To many, though, the rush for a patent—and profits—was a disturbing illustration of how Big Pharma was pushing boundaries.

"These are discoveries of nature, and it's baloney that we allow patents on living things," Jeremy Rifkin, a prominent antibiotechnology author, told the AP. "We didn't allow chemists to patent the periodic table, there's no patent on hydrogen, and I don't see why they can patent discoveries of nature."[3]

Since 1980, however, the United States government *has* allowed people to patent living things, as long as they are a new discovery, relevant to modern needs, and discovered using sophisticated scientific techniques. The legal argument was a twisted one. The CDC's patent covers the "isolated coronavirus genome, isolated coronavirus proteins, and isolated nucleic acid molecules."

According to John Doll, Director of Biotechnology for the patent office, that was enough to merit the application. "It must have a real-world utility and there has to be the hand of man involved," he said in an NBC News report in October 2003.[4] The isolated nucleic acid molecules represented the "hand of man," and the CDC was awarded the patent in 2004.

It's important to note that while COVID-19 is commonly referred to as being caused by "coronavirus," the term refers to a set of clinical symptoms, and there's no evidence linking the most recent SARS CoV-2 to a particular clinical expression of disease.

COVID-19 was not disclosed until 2020 (despite the 2012 Chinese data, captured in an August 2020 *New York Post* article, showing that bat guano miners had identical symptoms leading to the Wuhan Virus sample),[5] but other coronaviruses were known to scientists far earlier. It's analogous to a "species" in animals.

While that initial patent may have been well intended, that was only the beginning. "They actually filed patents on not only the virus, but they also filed patents on its detection and a kit to measure it," Dr. Martin explained. "Because of that CDC patent, they had the ability to control who was authorized and who is not authorized to make independent inquiries into coronavirus. You cannot look at the virus, you cannot measure it, you cannot develop a test kit for it without CDC authorization.

"By ultimately receiving the patents that constrained anyone from using it," he continued, "they had the means, they had the motive, and, most of all, they had the monetary gain from turning coronavirus from a pathogen to profit."

Who is responsible for protecting Americans from the virus? The same people who stand to make billions if that virus became a pandemic. While it's certainly a possibility that the people involved acted purely for personal and financial gain, it's not the only interpretation of this history.

The vast majority of employees and associates of the CDC, NIH, Big Pharma, and their numerous allies sincerely care about saving lives. The same can be said for the vast majority of medical professionals. Most would choose death before they would break their oath.

Yet again, what's important to take away from this history is the fact that nearly everyone involved with monitoring and solving the pandemic problem stood to make a lot of money because of their position at the wheel. It's so crucial to be aware of the conflicts of interest—especially because of what happened next.

Dr. Martin continued, "Somewhere between 2012 and 2013, something happened. The federal funding for research that was feeding into places like Harvard, Emory, University of North Carolina-Chapel Hill, that funding suddenly became impaired by something that happened at the NIH, where the NIH got this little tiny moment of clarity and said, 'I think something we're doing is wrong.'" In 2014, the NIH said gain-of-function research on coronavirus should be suspended.

Gain-of-function studies are studies that increase the transmissibility of a disease. Gain-of-function research is often necessary in order to study how a disease affects humans—and how to cure it. If a disease isn't strong enough to infect an animal or human cell at its most basic level, scientists must "teach" it to become infectious before they can study the aftermath.

Obviously, the danger is that once scientists teach a virus to infect humans, it can then escape and wreak havoc. Despite robust safety protocols, breaches continue to plague medical laboratories, and the risks to humanity are ever-present.

A statement released by NIH Director Francis S. Collins at the time read, "NIH has funded such studies because they help define the fundamental nature of human-pathogen interactions, enable the assessment of the pandemic potential of emerging infectious agents, and inform public health and preparedness efforts. These studies, however, also entail biosafety and biosecurity risks, which need to be understood better. NIH will be adhering to this funding pause until the robust and broad deliberative process described by the White House—including consultation with the National Science Advisory Board for Biosecurity (NSABB) and input from the National Research Council of the National Academies—is completed."[6]

However, there was a bit of small print that followed, and it made all the difference to coronavirus researchers. "During this pause, NIH will not provide *new* funding for any projects involving these experiments," Collins stated. Anyone "currently conducting this type of work" was encouraged to "voluntarily pause" their research until the government could make final recommendations, which was expected to take many years.

What to do in the meantime? Dr. Martin explained, "You off-shore the research. You fund the Wuhan Institute of Virology to do the stuff that sounds like it's getting a little edgy with respect to its morality and legality. But do you do it straightway? No, you run the money through a series of cover organizations to make it look like you're funding a US operation, which then subcontracts to the Wuhan Institute of Virology."

In April 2020, *Newsweek* magazine stated, "The NIH (with NIAID Director Dr. Anthony Fauci's backing) promised $7.4 million to the EcoHealth Alliance to study bat coronaviruses from 2014 to 2019—and in doing so, to conduct gain-of-function research. A large portion of that went to the Wuhan Institute of Virology. The lab also received millions from a program called PREDICT, funded by the United States Agency for International Development, which works closely with the NIH."[7]

Why send money from American taxpayers to a lab in China? To start, despite the ban, NIAID head Dr. Fauci was determined to see "gain-of-function" research continue. A huge proponent of such research, Fauci had already written a controversial op-ed for the *Washington Post* in 2011, arguing for the importance of his own gain-of-function research on the bird flu.

"Determining the molecular Achilles heel of these viruses can allow scientists to identify novel antiviral drug targets that could be used to prevent infection in those at risk or to better treat those who become infected," he wrote with two coauthors on December 30, 2011.

"Decades of experience tells us that disseminating information gained through biomedical research to legitimate scientists and health officials provides a critical foundation for generating appropriate countermeasures and, ultimately, protecting the public health."[8]

The Obama administration disagreed, and the ban came down in 2014, effectively ending that argument. In 2017, however, the NIH ended the ban under President Trump and fired up gain-of-function research stateside with one caveat: a secret panel of experts would weigh the risks and potential benefits and decide who would be allowed to proceed.

Scientists around the world were outraged to find that such delicate decisions were being made behind closed doors, especially when it was revealed that two risky flu studies—Fauci's area of focus—had been approved. In early 2019, Tom Inglesby of Johns Hopkins University and Marc Lipsitch of Harvard wrote a blistering op-ed for the *Washington Post*, ringing the alarm about the controversial research practices that had been fired up again.

"We have serious doubts about whether these experiments should be conducted at all," they wrote. "With deliberations kept behind closed doors, none of us will have the opportunity to understand how the government arrived at these decisions or to judge the rigor and integrity of that process."[9]

For Dr. Fauci and other gain-of-function researchers, China represented the ultimate closed door, Dr. Martin told Mikki. Despite the risks—and past investigations had proved that there were *plenty* at the Wuhan lab—millions of dollars continued to flow in. For the NIH bigwigs, that also minimized the potential risks to their reputations.

In the event of a security breach, Dr. Martin explained in his interview with Mikki, "The US could say, 'China did it.' China could say, 'The US did it.'" He joked, "The cool thing is, both of them are almost telling the truth. Because they did it together."

CHAPTER FIVE

The Gatekeepers

The media's the most powerful entity on earth. They have the power to make the innocent guilty and to make the guilty innocent, and that's power. Because they control the minds of the masses.

—Malcolm X

United States
March 2020

The speed of modern life and the omnipresent onslaught of information being pushed at us at any given moment makes it nearly impossible for people to fully research the events, people, policies, and decisions that shape their lives.

We may try to expose ourselves to multiple news sources in order to get a balanced look at what's true and what's not. If a story appears in multiple outlets, then it must be true, right? Wrong. While there are an infinite number of talking heads, most of them are reading the same script. The question is, who writes those scripts? In the words of Plato, "Those who tell the stories rule society."

In 2013, President Obama signed into law the National Defense Authorization Act (NDAA), which lifted restrictions on the domestic dissemination of government-funded media. In short, this amendment created loopholes that enabled media corporations to use propaganda against its own citizens.

On March 26 and 27, 2015, the *Institute of Medicine* convened a workshop in Washington, DC, to discuss how to achieve rapid and nimble MCM (medical countermeasures) capability for new and emerging threats. NIAID funding facilitator Peter Daszak lamented the absence of public support for funding vaccine development. He insisted that there needed to be a concerted propaganda campaign to coerce the public into getting behind universal vaccination. "To sustain the funding base beyond the crisis," he said, "we need to increase public understanding of the need for MCMs such as a pan-influenza or pan-coronavirus vaccine. A key driver is the media, and the economics follow the hype. We need to use that hype to our advantage to get to the real issues. Investors will respond if they see profit at the end of the process."[1]

For many people, Google is the onramp to the information super-highway. How can we even find information without going there first? Because of that fact—and because their original motto was "Don't Be Evil"—we might be forgiven for thinking that they are an unbiased platform. That couldn't be further from the truth. Google is the platform for upward of 90 percent of all online searches today. In 2019, their ad revenue was more than $139 billion. They wield that power like a bludgeon, and sometimes, people get hurt.

Robert Epstein, a Harvard PhD and former editor in chief of *Psychology Today*, launched a lifelong crusade against Google after his own experience watching his website be flagged for malware. In a 2013 *Time* magazine article, he blasted the tech giant's "fundamentally deceptive business model," and in 2015, he announced that he believed Google could work with other tech companies to rig the 2016 election.

In July of 2019, at a Congressional hearing, Senator Ted Cruz was aghast to hear Dr. Epstein's conclusion. "You testified before this committee. You said in subsequent elections, Google and Facebook and Twitter and big tech manipulation could manipulate as many as 15 million votes in a subsequent election," he bellowed.[2]

Dr. Epstein responded, "And the methods that they're using are invisible. They're subliminal. They're more powerful than any effects that I've ever seen in the behavioral sciences, and I've been in the behavioral sciences for almost 40 years."[3]

Dr. Epstein called these methods the "Search Engine Manipulation Effect," or Google's attempt to favor one political candidate over another in the search engine results. As a whole, he found—according to a study published in the *Proceedings of the National Academy in Sciences*—these methods could cause up to 80 percent of undecided voters to come to favor that candidate. Worst of all, the voters wouldn't even know why they changed their minds.

Now, Dr. Epstein was not saying that Google ever *had* or *would* rig an election. The main point is that they *could*, and that people should know about it. "That power exists, and, as long as it does, Google poses a serious threat to the democratic system of government," he wrote in the Huffington Post. "Google executives have more power over elections worldwide than any small group of individuals has ever had in the history of humankind."[4]

In February of 2021, a *TIME* article by Molly Ball offered a shocking detailed confession of how such forces conspired to affect the outcome of the 2020 presidential election:

> There was a conspiracy unfolding behind the scenes, one that both curtailed the protests and coordinated the resistance from CEOs. Both surprises were the result of an informal alliance between left-wing activists and business titans. . . . Their work touched every aspect of the election. They got states to change voting systems and laws and helped secure hundreds of millions in public and private funding. They fended off voter-suppression lawsuits, recruited armies of poll workers and got millions of people to vote by mail for the first time. . . . But it's massively important for the country to understand that it didn't happen accidentally. . . . That's why the participants want the secret history of the 2020 election told, even though it sounds like a paranoid fever dream—a well-funded cabal of powerful people, ranging across industries and ideologies, working together behind the scenes to influence perceptions, change rules and laws, steer media coverage and control the flow of information. They were not rigging the election; they were fortifying it.[5]

Altering the course of human history really would only take one disaffected Google employee, and over the years, there have been plenty.

In 2012, it was revealed that Google engineer Marius Milner wrote a string of code that enabled Google Street View cars to siphon the data of private WiFi networks as they drove the streets taking photos for Google Maps. According to the *New York Times*, "That data collection occurred from 2007 to 2010."[6] Google was fined, and court battles followed.

Also in 2012, Google was fined $22.5 million after an engineer hacked Apple's Safari network to allow Google to place ads on the browser without Apple's approval.

With American democracy hanging in the balance, Dr. Epstein argues, Google can't afford to make such mistakes. Yet, even if they manage to keep human interference at bay, the algorithm itself has the ability to skew the results. Then, there is the explicit censorship: the blacklist. Dr. Epstein has identified several different Google "blacklists," which have been corroborated by Google whistleblowers, hackers, and more.

Violating Google's Terms of Service agreement means that your account could be blacklisted. "But I would never do that!" you might be saying to yourself. You might be surprised. The problem is threefold.

First, most people don't actually read what's in the Google Terms of Service agreement. So, you could be flagrantly violating it without knowing.

Second, if you *do* suddenly find yourself locked out, good luck trying to find out *why*. Google is notorious for having a brick wall of a consumer service department, and it's highly unlikely that they'll take the time to explain their decision.

When this happens to small business owners, it can severely affect your company and even put you out of business. Banishment to the Siberia of the Internet can happen to individual websites, as well, which is equally crippling. If the Google robots find that your website is in violation of their closely guarded guidelines, you'll find yourself knocked back several pages in the search results, or banned altogether.

Google's tentacles reach far beyond its search platform and suite of Google products. YouTube is also owned by Google and has been home to some of the very worst censorship of all. The sheer volume of content on the site makes it impossible for Google to review each video, so censorship begins with the viewers, who are asked to tag inappropriate videos. From there, Google employees review the content and issue penalties according to secret guidelines.

For *PLANDEMIC: INDOCTORNATION,* Mikki spoke to a Google whistleblower, Zach Vorhies, who got a rare firsthand view of Google's secret blacklisting when he worked as an engineer at the company. "The blacklist is something that Google said didn't exist, and they testified to that under oath," he said. "Now me as an engineer, I just did a search on Google's internal search engine, and guess what I found. It had blacklisted search terms like 'cancer cures.' Why is Google deciding what people can and cannot search for?"

If Google is blindly trusted, then there are a handful of other sites that are trusted by virtue of their marketing. They present themselves as "unbiased," or "fact-checking," sites, when in reality they are anything but.

For example, Snopes.com. The husband-and-wife duo of David and Barbara Mikkelson founded Snopes.com in 1995. With no journalism background or training whatsoever, they built their fact-checking empire by using Google as their primary verifying source—and, by way of Google Ads, as their primary income source, too.

The Mikkelsons divorced in 2015. Barbara sued David for embezzling money that he allegedly spent on prostitutes—as well as a lavish honeymoon with his new wife, who worked as an escort in Las Vegas. Then, in 2017, David Mikkelson's new business partners filed a lawsuit accusing him of multiple counts of fraud and embezzlement.[7] Normally this would be a private matter, but when your website claims to be "the Internet's go-to source for discerning what is true and what is total nonsense," such information becomes a matter of public interest.

Despite all of these occurrences, people still trust them to hold this ethical position within the information landscape, and for them to be an unbiased news source. Yet even just one example illustrates how wrong that is.

Here's one of countless examples of top fact-checkers getting it wrong: In late January of 2019, Snopes—along with PolitiFact and factcheck.org—raced to squash the notion that coronavirus and its treatments had been patented. However, in doing so, they reviewed only 3 of the 4,452 publicly available patents, which unmistakably show that SARS coronavirus detection and treatment had been widely patented by both the public and private sectors.

More often than not, independent fact-checkers are neither independent nor factual. They're just as susceptible to the financial conflicts of interest, political biases, and authoritarian groupthink as their mainstream news allies. Simply put, they are political spin machines.

Let me stop now to reiterate: This is not a relativist worldview. There *is* truth and fact in this world, and we should all be searching for it. It's crucial to understand, however, that from the moment we open our eyes each day, our experience of reality is curated. Therefore, whether it's our news media, our government, or our family doctor, we need to remain constantly vigilant to the forces that are working hard to shape our perceptions and beliefs. Identifying and understanding these forces will help us decide what is true and what is false.

Historically, Americans have been susceptible to misinformation and full-on cover-ups from the medical establishment, in collusion with the government. Both are highly trusted entities, and for good reason. We depend on both to safeguard our lives. However, that blind trust is not warranted. There are endless examples of how doctors and politicians have colluded to lie to the American public.

Pointing it out in real time is treated as sacrilege. Those brave enough to speak up are too often mocked, shunned, and silenced. But in time, often many years later, truth prevails. As Hollywood turns these stories into movies or TV shows, the world is captivated, moved, and indignant. Collectively, we ask, "How could this have happened?" We insist, "As a society, we've come a long way from that." From *Dallas Buyers Club* to *Erin Brockovich*, *Dark Waters*, or *Spotlight*, the list goes on and on. The truth is often simply too much to handle when it's surrounding us. Time provides perspective and gives us the opportunity to process the information.

This will happen with the COVID-19 pandemic. Today, we are in the midst of the rapids, trying desperately to navigate the rocks and keep ourselves afloat. It's only when we reach calmer waters that we'll be able to look back and understand the route that got us there.

We've been through this before—many times, in fact—and not just with cigarettes, to name the most obvious example.

Take Agent Orange, or DDT, which were promoted as harmless household items for years before they were found to be devastatingly dangerous. In fact, the CDC advised Americans to use DDT in their homes. Ads claimed that it was "absolutely harmless" to both people and animals. Today, we know differently.

It took a while, but over time we became aware of DDT's horrific side effects. The CDC went back on their earlier pronouncements and warned that it could potentially cause cancer. DDT was banned in 1972.

Then there's the swine flu scare of 1976. When the disease hit a military base at Fort Dix in the winter of that year, there was some fear it could blossom into a pandemic that would rival the 1918 flu outbreak. US President Gerald Ford rushed into action with a wide-scale vaccination plan.

Straight from the White House in March 1976, he vowed to vaccinate "every man, woman, and child in the United States." The next month, he signed emergency legislation for the National Swine Flu Immunization Program and launched a full-scale PR campaign with celebrities and government officials doing photo ops at their vaccination appointment. (Sound familiar?)

No one stopped to question whether it was the right course of action, and in less than a year, 45 million people—roughly 25 percent of the US population at the time—were vaccinated. Meanwhile, concerning problems began to arise. People were getting sick—in most cases, sicker than the people who caught the actual swine flu. It soon became clear that the vaccine was doing more harm than the disease itself.

Unbelievably, only two people had to die from swine flu to trigger the creation of the program. Both were at the same military base, and one of them had underlying medical conditions.

Before long, the disease faded into a whimper, but the results of the dangerous vaccine lingered. More than 450 young people were diagnosed with the paralyzing disease Guillain-Barré syndrome, which was linked to the vaccine. CDC officials later admitted that they knew neurological side effects were possible, but they did not disclose that to the public when pushing the vaccine. This was nearly twenty years after the widespread American polio vaccine was mistakenly issued as a live virus, leaving 40,000 children with polio, 200 children paralyzed, and ten children dead. That also was done with the full awareness of the CDC.

To quote Dr. Martin in *PLANDEMIC: INDOCTORNATION*, "That CDC is the CDC that is now allegedly looking out for your public health."

The medical establishment has never been big on mea culpas. Likewise, news outlets are slow to print retractions that point out their own mistakes. Instead, they'd rather carry on with the new story, pretending as if the old one never existed.

As dozens of new articles are posted, they push down the old ones until it's hard to find proof that any other narrative ever existed. Then, in defiance of journalistic ethics, they censor and slander citizens for repeating the very claims they originally published.

Dr. Martin explained, "Every media outlet has planted evidence, and they have reranked pages. So if you look today at face mask wearing and if you look today at social distancing studies, you will see the studies that used to be number one, number two, number three on the pages of page rank search don't exist anymore. What is there, are studies that wind up having headlines that support the common narrative."

Even as governments flexed their muscles to restrict the freedom of its citizens, there was another class of men coming into creation that would be more powerful even than the president: the robber barons. In the late 19th century, a handful of savvy businessmen became incredibly wealthy by investing in oil, steel, and other natural resources. They also became notorious for their ruthless business practices, and monopolistic tendencies.

One such tycoon was John D. Rockefeller. By the turn of the twentieth century, Rockefeller controlled roughly 90 percent of the nation's oil reserves. As a result, he was the richest man alive during his time—and is even considered to have been the richest person *ever* to this day.

Once Rockefeller had scooped up almost all the oil in the country, he began searching for other ways to stay rich and increase his wealth. A great way to do this, he soon discovered, was by creating increased demand with new products based on his oil. The greatest market segment that he identified was medicine.

Kerosene, also known as "coal oil," became known as an antiseptic and all-purpose remedy at the turn of the century. It was recommended for ailments as diverse as scrapes and cuts, rheumatism, or a sore throat. Although the *Journal of American Medical Association* noted that some patients observed blistering and other unpleasant side effects, Standard Oil continued to push the product.

Rockefeller's approach was multipronged and truly unprecedented. Before his Standard Oil came on the scene, Americans relied on mostly natural remedies to treat illness. Shifting society to an allopathic model of disease care—that is, one dependent on synthetic drugs and surgeries—would take more than a clever ad campaign.

John D. Rockefeller crushed all competition by buying out pharmacies and forcing them to only carry what was later coined as "Rockefeller medicine." With nearly infinite financial resources,

Rockefeller dropped his prices so low that mom-and-pop pharmacies were shut out of business.

Rockefeller then snapped up every newspaper in sight and instructed them to sing the praises of his new line of drugs.

Rockefeller founded the Rockefeller Institute for Medical Research and installed his brother as the head. His mandate was to drive out all natural—and therefore unpatentable and unprofitable— medicines, creating a new market for oil-derived drugs.

Rockefeller shelled out hundreds of millions of dollars to transform his reputation from the most hated man in America to a benevolent philanthropist.

In 1910, Rockefeller hired educator and doctor Abraham Flexner to tour the nation's medical schools and submit his findings in a report. Ultimately released by the Carnegie Foundation, "Medical Education in the United States and Canada" was an analysis of medicine and medical education in the United States. A harsh critique, it essentially argued that there were too many doctors, too many medical schools, and too many natural medical techniques, marked as "quackery" being practiced.

The report led to the closure of many small, for-profit medical schools (i.e., the competition for the schools Rockefeller paid for). The number of schools in the country was cut in half between 1910 and 1944. Tragically, all but two of the African American medical colleges in the United States were shut down, largely due to Flexner's most racist "observations."

At the same time, Rockefeller gave $100 million to schools, hospitals, doctors, and scientists that would support his cause, through a shell foundation called the General Education Board. He helped prop up and steer the American Medical Association (AMA) as the ruling entity in charge of doctors' licensure.

Everything was set for Rockefeller medicine to become the only choice in America. Until signs began to arise that coal- and oil-derived medicines were causing cancer. In the face of bad press, though,

Rockefeller knew just what to do: give away more money. Rockefeller founded the American Cancer Society in 1912.

Following in the footsteps of John D. Rockefeller, today, the pharmaceutical industry spends at least twice the amount as big oil every year to influence laws, policies, and public perception. Thanks to Mr. Rockefeller, no industry has more power over our lives than Big Pharma.

Of course, Big Pharma does not include the little guys—the doctors, nurses, and medical professionals on the front line. More than anyone, they labor under the stranglehold of the forces above them. When crises like the COVID-19 pandemic emerge, they must fight to make their voices heard above the bellowing from their bosses, mainstream media, and the government.

Big Pharma also has a stranglehold on the media. A 2009 study by *Fairness and Accuracy in Reporting* found that every major media outlet in the United States (except for one) had a member of a major drug company on its board. Each year, Big Pharma "invests" about $5 billion in advertising with those media networks.[8]

While this may occur as a new issue for those just waking up to US media corruption, the fact is, the news narrative has long been compromised. As technology advanced and the dissemination of information was made possible, governments got involved. Most Americans don't realize that there was an official "Office of Censorship" operated by the federal government from 1941 to 1945.

Unofficial censorship, via the prudish Hays Code of the Motion Picture Producers and Distributors Association, occurred long before and long after that. The Code dissolved in 1956, but the censorship remained.

While Joseph McCarthy captivated Americans with claims of communist mind control and brainwashing in the media, a real network of media manipulation was taking shape behind the scenes in Operation Mockingbird. By 1953, CIA Director Allen Dulles had infiltrated roughly twenty-five newspapers and wire agencies, via four hundred sympathetic reporters. Carl Bernstein described the network in a 1977 article for *Rolling Stone* magazine:

> Some of these journalists' relationships with the Agency were tacit; some were explicit. There was cooperation, accommodation, and overlap. Journalists provided a full range of clandestine services—from simple intelligence gathering to serving as go-betweens with spies in communist countries. Reporters shared their notebooks with the CIA. Editors shared their staffs. Some of the journalists were Pulitzer Prize winners, distinguished reporters who considered themselves ambassadors without-portfolio for their country. Most were less exalted: foreign correspondents who found that their association with the Agency helped their work; stringers and freelancers who were as interested in the derring-do of the spy business as in filing articles; and, the smallest category, full-time CIA employees masquerading as journalists abroad. In many instances, CIA documents show, journalists were engaged to perform tasks for the CIA with the consent of the managements of America's leading news organizations.[9]

Former CIA officer John Stockwell painted a dire picture in an interview. "It goes beyond your wildest imagination," he said in a vintage clip within *PLANDEMIC: INDOCTORNATION*. "Setting up student organizations so they could draw radical students in. Five thousand university professors co-opted to help the CIA manipulate people's minds. Journalists in the US, including big-name journalists, co-opted to function routinely to help the CIA put out stories and biases to the world."

In January of 1975, Senator Frank Church led a new Senate committee formed to investigate government operations and potential abuses carried out by the CIA, the NSA, the FBI, and the IRS.

In April of 1976, the Church Committee conducted an investigation into the CIA's influence over both foreign and national news organizations. During the trial it was revealed that the CIA had over 3,000 television and radio executives, journalists, newspaper editors, and even book publishers under contract and under their control.

Under oath, the CIA was forced to admit that they were maintaining a global network through which they were manipulating public opinion through the use of state-funded propaganda.

After the trial, when asked about the controversial practices of Operation Mockingbird, and whether or not the program would continue, Sig Mickelson, the first head of CBS TV News, replied, "Well, yeah. I would think probably for a reporter, it would continue today." As featured in a vintage clip in *PLANDEMIC: INDOCTORNATION*, Mickelson continued, "But because of all of the revelations, I think you've got to be much more careful about it."

In 1981, during a staff meeting at the White House, CIA Director William Casey said out loud, "We'll know our disinformation program is complete when everything the American public believes is false." Today, a takes a simple scroll through the cognitive minefields of social media to confirm that Mr. Casey's program is indeed complete.

Mikki added, "Never before has our collective consciousness been so disconnected from reality. By design, we're losing touch with our innate ability to think critically and independently." He continued, "Through the consolidation of information, those who seek total control over humanity have been capturing our attention, literally, for generations."

"Let this sink in: Over 90 percent of *everything* you read, listen to, and watch is owned and controlled by roughly six corporate empires," Mikki explained. "Through this grand monopoly they create the illusion of truth. When TV watchers are flipping through channels and see the same news on what appears to be distinct and competing major networks, common sense says, 'it must be true.' But in reality, it's really just one story being uploaded to thousands of teleprompters from an undisclosed central location. That scripted information is then echoed around the world by a series of interchangeable talking heads. Talking heads that are paid 6 to 40 million dollars per year to lie to their own people and to subvert their own homeland," Mikki concluded.

During the COVID-19 lockdowns, life for Americans became like George Orwell's *1984*, where the screens were our only source of news and information. In the twenty-first century, Big Brother is certainly watching. More important, though, is the fact that Big Brother is writing—writing the story of the human race and simultaneously making sure it's distributed across all of our screens, 24/7/365.

Mikki expanded on this thought: "Science has proven time and again that stress and fear depletes our natural immune system, which is our first defense against viruses. Science has also proven that human-to-human connection is critical to our health and healing.

"Another well-known scientific fact is that the human body requires Nature to survive. Being outside among the natural world increases oxygen flow, reduces high blood pressure, and optimizes our heart, mind, and body."

He added, "Our Nation's remedy to COVID-19: Stay inside, far away from sunlight and all living things. Smother your oxygen intake by wearing a mask over your mask. Fill your home with toxic disinfectants. Tune into the doom preachers of corporate media and politics."

"If I didn't know any better, I might think some people actually want the death toll to increase."

While billions of lives have been devastated by the never-ending lockdowns, a *Forbes* article released April 2021, by staff writer Chase Peterson-Withorn, offers a perspective that explains why a select few are in no rush to reopen our world. "Twenty million Americans lost their job in the pandemic," Joe Biden remarked in his Wednesday night address to Congress. "At the same time, roughly 650 billionaires in

America saw their net worth increase by more than $1 trillion . . . and they're now worth more than $4 trillion."[10]

That's true, according to *Forbes*'s data—but the numbers are actually a bit richer. Total American billionaire wealth stands at $4.6 trillion as of the stock market close on April 28, 2021, by our count. That's up 35 percent from $3.4 trillion when markets opened on January 1, 2020, just as Covid-19 was beginning to take the world by storm.

Billionaires aren't just richer since the pandemic began; there are also more of them. A record 493 new faces joined *Forbes*'s World's Billionaires list this year—roughly one new billionaire every seventeen hours between March 2020 and March 2021—including 98 newcomers from the US that includes famous faces like Kim Kardashian West, moviemaker Tyler Perry, and Apple CEO Tim Cook.

"If we've learned anything from the COVID19 'plandemic,' it's that our world is increasingly controlled by technocrats," said Mikki.

tech·noc·ra·cy

noun

The government or control of society or industry by an elite of technical experts

Welcome to the new normal.

CHAPTER SIX

The Dress Rehearsal

The control of information is something the elite always does, particularly in a despotic form of government. Information, knowledge, is power. If you can control information, you can control people.

—Tom Clancy

October 18, 2019
The Pierre Hotel, New York

———

"On behalf of our center and our partners, the World Economic Forum and the Bill & Melinda Gates Foundation, I'd like to extend a very warm welcome to our audience here in New York, as well as our larger virtual audience participating online today. The goal of the Event 201 exercise is to illustrate the potential consequences of a pandemic and the kinds of societal and economic challenges it would pose."[1]

It's beyond macabre in retrospect. Leaders of multimillion-dollar companies and government agencies had gathered at one of Manhattan's most luxurious properties to playact their way through a global pandemic that killed thousands. One by one, they would opine at length about how *they* would handle such a global crisis, only to pat one another on the back at the end for saving the world—at least, for pretend.

"The Event 201 scenario is fictional, but it's based on public health principles, epidemiological modeling, and assessment of past outbreaks," the speaker explained. "In other words, we've created a pandemic that *could* realistically occur."

The simulation kicked off with a well-produced—but fake—news video. "It began in healthy-looking pigs," a polished female newscaster announced solemnly, over B-roll of a writhing herd. "Months, perhaps years ago. A new coronavirus spread silently. Infected people got a respiratory illness with symptoms ranging from mild, flu-like symptoms to severe pneumonia," the voiceover continued, as chilling images were projected on the screen at the front of the room. "The sickest required intensive care. Many died. At first, the spread was limited to those with close contacts . . . but now it's spreading rapidly throughout local communities." International travel helped the illness hop borders, the news reel explained, until it was a full-scale global pandemic.

The simulation predicted the spread of conspiracy theories, as the elite panel discussed the most effective ways to prevent the flow of public disinformation. Censorship was rampant as millions clamored for a vaccine—even one that would be experimental and not fully tested. Hospitals were overflowing, and masks and gloves were scarce.

Event 201 took place in October 2019—five months before COVID-19 was declared a pandemic. An event of this complexity and magnitude would take months to write, prep, and produce, placing its date of conceptualization at least one year prior to the actual pandemic.

The question that arises for anyone paying close enough attention is—if this collection of wealthy and powerful knew that far in advance exactly what would be needed and in short supply, why did they wait till the actual pandemic to begin addressing those critical details?

Mysteriously, while Event 201[2] was "hosted" by Johns Hopkins University, the World Economic Forum, and the Bill and Melinda Gates[3] Foundation, it was paid for by Open Philanthropy, an opaque charity run by Facebook cofounder Dustin Moskovitz. An investor in Chinese CRISPR technology company Sherlock Biosciences, Dustin had considerable gain from an "epidemic" that would get his technology authorized under an Emergency Use Authorization (EUA).[4] (In fact, that's what wound up happening.)

All of the participants described the woeful lack of Personal Protection Equipment (PPE) and other disaster resources, but Gizmodo zeroed in on what was perhaps an even bigger threat: censorship and misinformation regarding the pandemic. "Social media outlets also fanned the flames by allowing trolls and even governments to spread disinformation about [the fictitious virus], such as blaming foreigners for the problem," author Ed Cara wrote. "That in turn made people even less likely to trust public health experts."[5]

Eric Toner, project director of Event 201, told Cara, "We're seeing right now, with recent outbreaks like Ebola, that social media plays a big role, both positively and negatively. It's how many people get their news now, but it's also how rumors and misinformation get propagated."

For the power players, that was a crucial takeaway. Even more than stockpiling PPE—which they didn't ever do—was ensuring control of the sources of information. Lockdowns would make that ever more vital.

Dr. Martin explained, "If you can keep people from assembling, guess what they're not talking about. They're not talking about the issues of the campaign. If you can keep people in their homes, the only source of information that you can have is what you curate for them.

"Now I know how to target my electorate," he continued. "They're in the only place I allow them to be, being fed the only message I'm allowing them to hear, through a media that I control."

On March 2nd, 2021, Zosimo T. Literatus of Yahoo News did what journalists used to do—research. His journey into the rabbit hole began with a simple question.

"Last year, the documentary 'Plandemic: Indoctornation' had hit social media, facing criticisms of falsehood and outright attacks from the print and digital media. Its arguments were convincing and evidence apparently strong. Some critiques even claimed it as disinformation. Of course, the critiques could be true. However, what if they were wrong?"[6]

Mr. Literatus followed up with a series of reports on Plandemic: Indoctornation, April 20th, May 4th, and June 1st of 2021. To his credit, he personally reviewed the key patents and points revealed by Dr. Martin. In the end, Mr. Literatus was unable to find anything that justified the "debunked" label the movie had been given by critics.

In his June 1st article, Zosimo T. Literatus wrote:

> Most media outlets labelled the film content as a "conspiracy theory," which supposedly promotes misinformation particularly of the Covid-19 pandemic. What falls flat in the claim of "conspiracy theory" and "misinformation" is the mysterious and the impeccable timing of the October 18, 2019 scripted simulation in Event 201, which described closely the Covid-19 pandemic that erupted in China in December 2019 three months later . . .
>
> It must be noted that the China declaration of the Covid-19 breakout could have been delayed by state censorship until the event could not be hidden from the world in December 2019. This means that the breakout could have occurred a few months earlier. The big question that was never answered: Did the convenors

of the October 2019 Event 201 know what was happening in China long before the simulation event took place?[7]

Event 201 wasn't, however, the only exercise of its kind. In the run-up to coronavirus, scores of the world elite were drafting various versions of the same event.

The "World at Risk Scenario" was released a month before Event 201, in September 2019. The "World at Risk Scenario" was drafted by an organization called the Global Preparedness Monitoring Board, which is a part of the World Health Organization. Board members for the GPMB include Dr. Anthony Fauci; Dr. Christopher J. Elias, the president of the Global Development Division of the Bill & Melinda Gates Foundation; and George Fu Gao, the director of the nonprofit Chinese Centers for Disease Control and Prevention, which is largely funded by the American CDC.

According to the WHO, the report "provided a snapshot of the world's ability to prevent and contain a serious global health threat." More important, the group recommended "seven urgent priority actions leaders must take to prepare across five areas," the WHO explained, including "leadership, building multi-sectoral country systems, research and development, financing, and robust international coordination."[8]

Specifically, they recommended that by September 2020, global leaders should complete two global pandemic preparedness exercises, such as Event 201. One, they noted, should be focused on the release of a respiratory pathogen.

Back in January 2017, Dr. Fauci had issued a warning to the sitting president. In a pandemic preparedness forum at Georgetown, he said there was "no doubt" Trump would face a "surprise" pandemic before the end of his term.[9] Indeed, just over three years later, Dr. Fauci would preside over one.

Dr. Martin explained how he and his research team noticed the early signs that the world was preparing for a pandemic event—and that they seemed to know just what disease would cause it. "My systems flagged anomalies when I started seeing nonprofits and corporations and cover financing for coronavirus programs in the late summer and fall of 2019," he explained. "Our first red flag came out when we read the World at Risk scenario," in September 2019.

As he watched, the story they had drafted began to come true. "Every single thing that you have seen play out in front of your eyes, all of them laid out in their tabletop exercise," he said. Was it really just a coincidence?

Dr. Martin was stunned by the synergies, which "fact-checkers have said has nothing to do with the coronavirus outbreak," he said. "Just happenstance. This is that wonderful universe of improbabilities where events just coemerge and then nature conveniently backs itself into our architecture. That's the scenario we're supposed to accept. Brilliant."

One year prior to Event 201, the Johns Hopkins Center for Health Security brought together many of the same sponsors, hosts, and actors to produce a tabletop pandemic simulation for a fictional virus they branded Clade X. At the time, that was the third pandemic play that they had put on. The first, Dark Winter, was held in 2001, just a few months before 9/11.

"Its timing, just a few months before 9/11, made its terrifying outcome—the near-complete breakdown of government and civil society—deeply resonant," Nicola Twilley, a reporter for the *New Yorker* who watched the exercise, wrote. "Dark Winter is credited, in part, with spurring George W. Bush to pass Directive 51, a largely classified plan to ensure the continuity of government in the event of a 'catastrophic emergency.'"[10]

Though much of that document was classified, the sections that were made public suggested that the president could assume nearly authoritarian powers, including instituting martial law. In execution, we now know, it included warrantless wiretapping of American citizens, torture of perceived enemies, and more.

Authoritarian response overall was the outcome of a pandemic preparedness exercise released way back in 2010 by the Rockefeller Foundation. The 54-page document, called "Scenarios for the Future of Technology and International Development," features the pandemic scenario Lock Step, "a world of tighter top-down government control and authoritarian leadership with limited innovation and growing citizen pushback."

In the Clade X exercise, they explored how that authoritarian leadership could actually exert itself at the level of the states. As the exercise unfolded, governors imposed lockdowns and travel bans on Americans from other states—just like what happened with the real COVID-19.

In the exercise, as in reality, there was confusion as different states set different guidelines and the federal government declined to intercede. "In a serious outbreak, there will be federalism issues," Tom Inglesby, then the Director of the Center for Health Security at the Johns Hopkins Bloomberg School of Public Health, said. "They may be manageable. They may not be."

According to the *New Yorker's* Twilley, "Even within the artificial confines of the simulation, there was a lack of leadership. Everyone agreed that the president had the final word on everything in general, but nobody seemed to be responsible for America's outbreak response specifically."[11]

Politician Tom Daschle, a former senator from South Dakota who was playing the Senate Majority Leader during the exercise, complained, "We're five months into this crisis, and I still can't tell you who's in charge."

Twilley wrote, "Ironically, just a couple of days earlier, the person to whom this responsibility would have fallen in real life, Rear Admiral Tim Ziemer, had been removed from his position as the head of global health security on the White House's National Security Council . . ."[12] This all but ensured that the pandemic fantasy of Clade X chaos would turn into reality, allowing Big Pharma and big business to step into the power vacuum that the government had created.

Indeed, by the time that the COVID-19 outbreak flared up, Ziemer had not been replaced. With a lack of leadership at a federal level, and state governments focused on corralling their own citizens, who could oversee the global response to a medical disaster?

The World Health Organization (WHO) was established in 1948 specifically for that cause. A specialized agency of the United Nations, its stated goal is "the attainment of all peoples of the highest possible health." In practice, that looks a lot like what the CDC does in the United States: issuing guidelines, tracking epidemics, and keeping a pulse on public health.

In reality, it faces the same conflicts of interest as many other organizations, because it needs outside funds to function. Although the WHO is the institution granted exclusive power to guide and protect the health and wellness of humanity, it is sustained by private donations, the bulk of which are made by pharmaceutical and biotech corporations who have a vested financial interest in the organization's support.

For the 2018/2019 budget cycle, for example, the biggest volunteer donations from organizations (as opposed to the fees assessed from member states) were as follows:

- The Bill & Melinda Gates Foundation: $531 million
- GAVI (Global Alliance for Vaccines and Immunization, founded by Bill & Melinda Gates): $371 million
- Rotary International (a partner of the Bill & Melinda Gates Foundation for more than a decade): $143 million
- The World Bank (a partner of the Bill & Melinda Gates Foundation since 2018): $133 million
- The European Commission (the executive branch of the EU, which received over $45 million in donations from the Bill & Melinda Gates Foundation in 2019): $131 million
- National Philanthropic Trust (recipient of millions of dollars in grants from the Bill & Melinda Gates Foundation): $108 million[13]

As early as 2017, the world was starting to take notice of how Bill Gates—a man, as noted earlier, with no medical training—was beginning to insert himself into the world's most powerful international health organizations.

That year, on the eve of the appointment of a new WHO Director, Politico reported, "The software mogul's sway over the World Health Organization spurs criticism about misplaced priorities and undue influence."[14]

One of the candidates for the Director position was former Ethiopian health minister Tedros Ghebreyesus. If elected, he would be the first director of the WHO who wasn't also a doctor.

"Last year, a French diplomat suggested that Gates also supports Tedros," Politico added, "having funded health programs in his country when he was health minister."

The Gates Foundation denied it, telling Politico that "the foundation cannot take a position given that it is not a voting member country and thus has to remain neutral."

Still, was it any surprise that Tedros emerged victorious? With Gates behind him—as well as other powerful allies including the Clinton Global Initiative and the Chinese Communist Party—he was a shoo-in. The fact that he wasn't even a doctor was easily ignored. More disturbingly, however, were some of the scandals in his past that were quietly swept under the rug.

Prior to his appointment, Tedros was a high-ranking member of the Tigray People's Liberation Front in Ethiopia, a brutal and corrupt political group responsible for crimes against humanity, including bombings, kidnappings, tortures, and killings. He also was accused of helping to cover up three different cholera epidemics in the African nation that occurred under his leadership as health minister.

When the first cholera outbreak occurred, the disease was reclassified as "Acute Watery Diarrhea," or AWD. Critics said this was done to downplay the outbreak and prevent international concern. In actuality, it did little more than prevent the flow of foreign aid, and the implementation of a vaccine, both of which would have required the government to classify the disease specifically as "cholera." Ultimately, critics said, his citizens died to prevent his own embarrassment.

Aghast, a group of American doctors wrote a letter to Tedros pleading him to reconsider in 2017. "Your silence about what is clearly a massive cholera epidemic in Sudan daily becomes more reprehensible," they wrote. "The inevitable history that will be written of this cholera epidemic will surely cast you in an unforgiving light." As reported in May 2020 by the *Eurasia Review*, they blasted him for being "fully complicit in the terrible suffering and dying that continues to spread in East Africa."[15]

None of that made it to the résumé that Tedros submitted to the WHO for his consideration. To this day, he is still known in some circles as "Dr. Cover-Up," a man who, one think tank claimed, "allegedly personifies the erosion in impartiality and rectitude expected from such a storied body."

Dr. Martin explained, "We are living in a time where leadership, unfortunately, is compromised. By that I mean that individuals are placed in power for their ability to be influenced, not their merits of leadership. Nothing could be clearer than the leadership of the World Health Organization."

The true "leaders" of the world are the ones doing the influencing, the ones with the money to fund such organizations: a group that David calls "the interlocking directorates of the World Health Organization, the CDC, the NIAID, or the organizations that are the philanthropic cover organizations that fund them." Such "philanthropic cover organizations" include GAVI, the Bill & Melinda Gates Foundation's pro-vaccine spin-off, for example.

Presently, there are more than 1,300 patents issued and held by organizations that are either recipients of funding from or otherwise linked to those groups. It doesn't take much digging to find the fingerprints of Bill Gates and Anthony Fauci on almost all medical research and innovation today.

There's Sherlock BioSciences, for example, which creates diagnostic tests and is heavily funded by the Gates Foundation; and EcoHealth Alliance, a former partner of the Wuhan Virology Lab. The Alliance received funding from the NIH and had a director of the Gates Foundation on its board of scientific advisors. Moderna, one of the companies that led the push for the COVID-19 vaccine, has received millions in grants from the Gates Foundation and partnered with the NIH and NIAID for their vaccine trials.

It's a very incestuous web. Even if we don't believe that there is malice in the minds of those pulling the strings, we should at least be aware of how all of those strings are tangled. For most of our history, we actively tried to prevent such massive intermingling of corporate power.

Indeed, there was a time when pharmaceutical CEOs could be jailed for knowingly selling a dangerous product or covering up the evidence of its negative effects on public health. Today, they simply get a slap on the wrist in the form of a fine.

In just one recent—and egregious—example, see the Sacklers, the family behind Purdue Pharma, the makers of OxyContin. Largely credited with causing the American opioid epidemic that killed nearly half a million people, the Sacklers are alleged to have knowingly pushed their highly addictive drug simply to benefit their bottom line.

After extensive public outrage, Purdue Pharma pleaded guilty to criminal charges related to OxyContin. Instead of jail time for executives, Purdue Pharma was hit with a seven-figure fine. The Sacklers themselves agreed to pay $225 million: roughly one-quarter of one year's revenue from the drug, and 8 percent of the company's yearly revenue overall. That works out to about $450 per American who died during the height of the opioid epidemic.

For many, fines are just the cost of doing business in the pharmaceutical industry. According to Dr. Martin, however, it's much more serious. "If you have conflicts of interest in the funding and in the decision-making and in the inside knowledge that you have between competing or competitor organizations, that is a violation of the antitrust laws of the United States," he said. "These are federal crimes."

CHAPTER SEVEN

Gates of Hell

The welfare state is not really about the welfare of the masses.
It is about the egos of the elites.

—Thomas Sowell

According to legend, young Bill Gates was a computer geek who built his computer empire out of his Seattle garage. An underdog in a sweater vest and glasses, he somehow managed to "aw shucks" his way into billions of dollars and untold power and privilege. It's the ultimate American success story, right? Well, it is—but that's not the actual story.

Bill Gates actually was born into wealth and privilege. Both his grandfather and great-grandfather were banking moguls. His father, William Gates, Sr., was a prominent Seattle-based lawyer and political lobbyist. His mother was a businesswoman and political activist, who became the first woman to head the United Way, an organization that has been at the center of financial scandals since the early '90s. Clearly, these were people who knew the ropes when it came to power in America.

Bill went to Lakeside Prep School, then went on to Harvard but dropped out during his sophomore year to start the company that would become Microsoft. Even then, he wasn't just relying on his entrepreneurial spirit to make his way in the world.

With his fledgling company struggling, Bill's mom spoke to United Way Board Member John Opel, who also happened to be the chairman of International Business Machines, or IBM. According to Gates's mother's obituary in the *New York Times*, "Mr. Opel, by some accounts, mentioned Mrs. Gates to other I.B.M. executives. A few weeks later, I.B.M. took a chance by hiring Microsoft, then a small software firm, to develop an operating system for its first personal computer" in 1980.[1]

That was the boost that Bill needed. The IBM PC launch bundled MS-DOS (Microsoft Disk Operating System) into its products, launching Microsoft into the public consciousness as a major player in tech and starting Bill on the path to success. MS-DOS wasn't exactly his invention, though.

In December 1980, Microsoft paid $25,000 for the rights to market an operating system developed by Seattle Computer Products to other manufacturers. In July 1981, one month before the launch of the IBM PC, they paid an additional $50,000 for the full rights to the system. Microsoft allowed Seattle Computer Products (SCP) to continue selling their own OS, but without the hardware to put it on, the small computing company struggled.

Soon, the twenty-something owner of the company realized that his only remaining asset was that license to sell the software that was essentially Windows. As you might expect, there was widespread interest from Microsoft competitors who wanted to snap up the software at the heart of the Gates empire. Microsoft wouldn't stand for it.

Tensions escalated, and both sides became embroiled in a multi-million-dollar lawsuit that lasted for years. Eventually it was settled out of court, and Microsoft paid $925,000 to SCP—both for the right to sell the software that SCP invented and, essentially, for the right to claim that Bill invented it himself.

It was a mistake that Bill would never make again. Moving forward with Microsoft, he would become one of the most prolific forces in the patent system, filing tens of thousands of patents under countless holding companies. It would be both a great means of ensuring asset protection in the event of any eventual lawsuit and also a means of obscuring just how powerful Microsoft was becoming.

Dr. Martin explained in *PLANDEMIC: INDOCTORNATION* how it's become nearly impossible to truly unravel and trace the impact of Microsoft and Bill Gates on Big Pharma and other fields of innovation. "Part of the problem, and this, by the way, goes back to the 1980s with Microsoft, is Bill Gates found out that it was very difficult to manufacture a way to navigate through the patent universe . . .," he explained. "He became the architect of a very cunning program of putting patents in holding companies that didn't have anything to do with the name of the organization," thereby hiding the true extent and nature of his work.

In 2013, Microsoft briefly published a searchable list of more than 40,000 patents owned by the company and its subsidiaries. Today, that link no longer works. Instead, there is a page with generic corporate speak about the importance of patents.

In building his company, Bill Gates pulled a page from John D. Rockefeller's book. Flush with cash from his IBM deal, he could afford to cut deals and lower prices to levels that would crush his competitors—and make his own software the "gold standard."

Gates earned a reputation for being ruthless. Through his father, he had learned the ins and outs of law and politics and how to fight dirty to manipulate those governing forces. Even his closest friends were not immune.

Gates had founded Microsoft with high school buddy Paul Allen in 1975. Allen had been the company's tech chief but clearly found himself in over his head as development at the company took off. Disappointed with Allen's output, Gates recruited a fellow Harvard brainiac, Steve Ballmer, to join the company.

According to Allen's autobiography, *Idea Man*, he had told Gates that he'd be comfortable with giving Ballmer a 5 percent equity stake. Later, however, he found a document proving that Gates had given Ballmer more than 8 percent. The crowning indignity, though, was when Allen overheard Gates and Ballmer hatching a plan to dilute his shares.

He wrote, "They were bemoaning my recent lack of production and discussing how they might dilute my Microsoft equity by issuing options to themselves and other shareholders."[2] Allen's lack of production was not due to laziness. It was due to intensive chemotherapy, as he had just been diagnosed with non-Hodgkin's lymphoma. In a time of uncertainty and need for support, his life-long friend and business partner decided to use that specific moment of weakness and instability as a business opportunity to remove him from the organization that he'd helped build. Allen resigned from Microsoft shortly after and neither returned nor ever spoke to Gates again.

Today, most Americans think of Bill Gates as a sole founder. Allen has been edited out of the narrative. Gates was making some very big moves in the early years of Microsoft—and doing it without the blessing of the higher-ups in the American power structure. He was playing outside of the lines, and before long, he got smacked down for it.

In 1998, the US Justice Department and attorneys general in twenty states announced antitrust charges against Microsoft. They were trying to determine if Microsoft broke the law by bundling software and making it impossible to download competing programs on Microsoft hardware. Microsoft lost the case. A judge found that the computer company had violated the Sherman Antitrust Act and that they were operating a monopoly on the computing industry.

Microsoft was quick to appeal. The case was overturned. Still, the whole debacle didn't feel much like a win for Microsoft. Throughout the course of the trial, Gates gave eighteen hours of videotaped testimony, and the picture that emerged was not flattering. It challenged the rags-to-riches narrative of a Washington computer geek winning big.

From a bespectacled wunderkind, he'd evolved in the public consciousness into an evil billionaire who crushed bright young would-be entrepreneurs. All of a sudden, he was very easy to dislike. If he were to continue doing business in the public sphere, Bill Gates needed an urgent rebranding.

Overnight, he transformed his public image from ruthless tech monopolizer to the world's most generous philanthropist. Gates founded the Bill & Melinda Gates Foundation, and announced the Foundation's first $100 million seed donation in 2000. Over the next twenty years, the Gates Foundation would donate millions to health-care organizations, governmental organizations, and biotech companies around the world.

All of that money wasn't just coming from Bill and Melinda Gates's pockets, though. The Gates Foundation maintains a robust portfolio of investments in the very kinds of companies that they are support-ing with their cash, such as Merck, Lilly, Pfizer, and other Big Pharma giants. That means that their "grants" turn into savvy investments.

In essence, the Gates Foundation expanded rapidly into a massive, vertically integrated multinational corporation, controlling every step in a vaccine supply chain that reaches from its Seattle-based boardrooms to the villages of Africa and Asia.

Oddly, few people stop to question why a computer geek and col-lege dropout should be the world's self-appointed leading expert on medicine—and vaccines, in particular. No one blinks an eye when he gives speeches, interviews, and other statements about his view for the future of vaccines. He was even the first private individual (not to mention nondoctor) to keynote the WHO's assembly of member states.

One might think, *If he's willing to share his money to benefit the world, it doesn't really matter, right?*

Well first, there's a tangible financial return on his Pharma investments. More important is the influence such an investment conveys—influence to steer policy, and to have a seat at the table. Second, while we're all well aware of his "positive impact" thanks to the never-ending press releases and public appearances, the harm the Foundation has done in countries around the world—solid proof that he should leave medicine to the medical experts—frequently gets swept under the rug.

For just one example, look to India. In 2009, the Gates Foundation incentivized Bollywood stars and other celebrities to promote a new schedule of mass vaccinations. Over 24,000 girls—mostly from tribal communities—were given what they believed to be "wellness shots," often without the consent of their parents. Instead, they were allegedly untested HPV vaccines, administered by an organization called PATH (Program for Appropriate Technology in Health).

Mary Holland, Vice Chair & General Counsel, Children's Health Defense, explained in a Zoom call with Mikki: "The people that were administering these vaccines lied to the guardians of these girls and told the girls, 'Oh, this is going to cure cancer. You're never going to have cancer,' and these girls became severely injured. Some of them developed seizures, some of them developed cancer. Seven girls died, and there was no insurance, there was no assistance for them. The Gates Foundation denied that it had been clinical trials. It was so bad that the parliament in India created a task force."

Dr. Colin Gonsalves was on the Supreme Court of India when this investigation occurred. During a Zoom call with Mikki, he recalled the heartbreaking tragedy that was largely absent from American news.

"India is a barbaric country. Things happen here in a very barbaric way," he began. "But I was surprised to find an American organization operating in broad daylight, doing things in a very, very, let's say, Indian fashion."

When concerns arose regarding the Gates-funded vaccination program, "I wanted the whole procedure to be investigated," he explained.
"The Indian parliament formed a committee, and it was to me a rather
surprising move because you generally don't often have such a high-level
inquiry into matters affecting poor people."

At first, the results were predictable. The government determined
that the deaths of the seven girls were not all related to the vaccine:
"One girl drowned in a quarry; another died from a snake bite; two
committed suicide by ingesting pesticides; and one died from complications of malaria. The causes of death for the other two girls were less
certain: one possibly from pyrexia, or high fever, and a second from a
suspected cerebral hemorrhage," the report read. However, additional
investigation turned up disturbing issues.

In 2010, the government found that there had been ethical viola-
tions regarding informed consent. A 2011 report found that the study
had not created any kind of system for monitoring adverse effects.
Finally, in 2013, yet another report was issued, and it came down hard
on PATH and its partners.

"That was such an extraordinary report," Dr. Gonsalves remembered.
"I don't think the Indian parliament has ever come out with such a scathing report. The government officials came up and said, 'We shouldn't
have authorized this. We're sorry. We're not going to allow them again.'"

Science magazine reported, "Rather than endeavoring to protect
women's health, PATH, it charged, was a willing tool of foreign drug
companies hoping to convince the Indian government to include the
HPV vaccine in its universal vaccine program, a roster of mandatory
immunizations that the government is required to pay for." (HPV vaccine
continues to be available in the Indian private sector.)[3]

In particular, the panel saved its sharpest critique for the Indian
Council of Medical research, which it claimed had "completely failed
to perform [its] mandated role and responsibility as the apex body for
medical research in the country. . . . Rather, in [its] over-enthusiasm

to act as a willing facilitator of the machinations of PATH, [it has] even transgressed into the domain of other agencies, which deserves the strongest condemnation and strictest action against [it]."[4]

The report was a total humiliation for the Indian government, although PATH and certainly the Gates Foundation received little public blowback. Still, in 2017, the Indian government announced that they were cutting ties with the Gates Foundation in order to fund the country's Immunization Technical Support Unit themselves.

"Critics have in the past raised concerns [that] the foundation should not have any association with the program due to apparent conflicts of interest," Reuters reported at the time. "That's because the foundation also backs GAVI, a global vaccine alliance that counts big pharmaceutical companies as its partners."[5]

Senior Health Ministry Official Soumya Swaminathan told *Reuters*, "There was a perception that an external agency is funding it, so there could be influence."[6]

That certainly didn't mean that the Gates Foundation was gone from India for good. As recently as March 2020, the Gates Foundation was making strategic interventions into the country's medical establishment. That month, they made a sizable donation to the All India Institute of Medical Sciences "to help highlight the work being supported by Foundation in India and also catalyzes thinking and debate amongst the various levels of policy and decision-makers countrywide."[7]

For Dr. Gonsalves, sadly, it was déjà vu all over again. "And now they're back doing their same old tricks again," he said, shaking his head. "You can imagine . . . the manipulation of the media, the manipulation of public opinion by leaders from all political parties, unanimously saying, 'We want a vaccine.'"

With the coronavirus pandemic stoking worldwide demand for a vaccine, he's sure that the Gates Foundation is not done with India just yet.

Indeed, by the end of 2020, the Gates Foundation and their partners, such as Pfizer and Moderna, would be first out of the gate with supposed solutions.

"The worst thing is they are taken as philanthropists, whereas what this actually is is the acquisition of political and financial power," Dr. Gonsalves said.

While the Gates Foundation spins their work in India as some kind of noblesse oblige to a third-world country, Dr. Gonsalves is more pragmatic: "I think the second most populous country, with 1.3 billion people, is going to be a good base for pharmaceutical companies to make a killing. And also kill a lot of people in the process. It's so terrifying as to what they're actually doing with the world."

In 1986, Congress passed the National Childhood Vaccine Injury Act, and President Ronald Reagan signed it into law. The legislation was sweeping, completely restructuring the nature of vaccines in the United States. First, the NCVIA created the National Vaccine Program Office to coordinate all things vaccine-related across the CDC, FDA, NIH, and other governmental agencies. It also created a committee at the Institute of Medicine (a nonprofit organization) to review cases of adverse events following vaccination. Doctors were told to report such events to the new Vaccine Adverse Event Reporting System, comanaged by the CDC and the FDA. Finally, it tasked the CDC with developing Vaccine Injury Statements outlining the potential risks of every vaccine. Doctors would be required to provide these to all patients prior to vaccination, to ensure informed consent. (Although, they couldn't force them to read that fine print.)

If and when adverse events occurred, however, patients would no longer be able to sue the vaccine manufacturers directly. Instead, the NCVIA created a system called the National Vaccine Compensation Program to handle complaints. Anyone who believes that they were harmed by a covered vaccine (from the NVCP's list) can file a petition with the program and potentially receive compensation. That money doesn't come from the vaccine manufacturers, though. It comes from you, the taxpayer.

Every time someone gets a vaccine, they pay a $0.75 excise tax that goes straight into the "Vaccine Injury Compensation Trust Fund." With every child who is inoculated in America, that adds up quickly. From 2013 to 2017, for example, the program paid out an average of $229 million each year to American families. More than $4 billion has been paid out since its inception in the 1980s. That number represents a lot of injuries.

However, this number is not based on the number of injuries overall. Instead, it represents the small percentage of reported injuries that were able to go to court in search of damages. Not all vaccines qualify for the compensation program, and many people simply do not have the time or funds to go through the claims process.

The NCVIA is not the only factor when it comes to Big Pharma and COVID, however. The Vaccine Compensation program and the liability that came along with it got a major boost after 9/11, when the Public Readiness and Emergency Preparation Act (also known as the PREP Act) was signed into law. According to the Act, the Secretary of Health & Human Services can issue blanket liability immunity at any time to anyone involved in the development of countermeasures to diseases and other threats that are a risk to public health. Of course, they did that for COVID.

In March 2020, the HHS released a declaration for COVID-19, protecting literally any person, company, or organization involved in developing, manufacturing, testing, distributing, or administering anything related to COVID-19. No matter what happens, as long as it isn't "willful malpractice," no one in Big Pharma can ever be sued for anything related to COVID-19. If you take the vaccine and get sick, you are on your own. If you wear a mask and it makes you sick, too bad. If you lose your job because you get sick, or if your health insurance goes through the roof because of it, tough luck.

Vaccines *can* work. That's science. They *can* save lives. Still, is it really a good idea that we've created a system whereby vaccine manufacturers have literally *no* incentive to make sure that their products are safe? Yes, we want to assume that they actually care about whether they hurt people, but we know from history that good intentions aren't always enough to keep people from getting killed.

At the same time, there's huge incentive to be first to market. Even before coronavirus hit, there was a massive shift in the medical field toward more vaccine research. One study in *European Molecular Biology Organization Reports* found that funding for general vaccine-related research increased by 41 percent from 2003 to 2007, with funding for malaria and tuberculosis vaccines in particular increasing by 96 percent and 62 percent, respectively.

Meanwhile, other disease research took a hit. For example, funding for research into heart disease—still the leading cause of death in the US—increased only 3 percent during that time.

The EMBO researchers were definitive: According to the report, the interest in and financial support for global health initiatives like vaccines at the NIH was pretty sudden, and it was almost entirely due to the influence of the Bill & Melinda Gates Foundation. They didn't go so far as to say that the Gates Foundation *told* the NIH to push for vaccines. However, they did point out that politicians are much more susceptible to pressure from outside groups. Congress has no problem telling the NIH where to put their funding. That's what happened with AIDS in the 1990s, or breast cancer research more recently.

Essentially, Bill Gates is setting the medical agenda for our country—and with that, the world. The fact that the public mission of the Gates Foundation is to improve and protect the health of all people, yet its founders are major investors in an array of companies known for being the worst polluters of our planet and our bodies, should raise questions.

In January of 2007, Amy Goodman of *Democracy Now* asked, "Is the world's largest private philanthropy causing harm with the same money it uses to do good? That's the question hanging over the charity of Microsoft founder Bill Gates and his wife Melinda today. The *Los Angeles Times* has revealed the Bill and Melinda Gates Foundation has made millions of dollars each year from companies blamed for many of the same social and health problems the foundation seeks to address."[8]

That was back in 2007. Since then, Mr. Gates has grossly expanded his global monopoly. He is not only one of the largest stakeholders in Monsanto, the developer of Roundup, an herbicide proven to cause cancer, but Gates also has his hands in just about every major service company we depend on. Giants such as Apple, Amazon, United Parcel Service, FedEx, Crown Castle International (real estate), Canadian National Railway Co., Caterpillar, Waste Management Inc., Berkshire Hathaway (holding company), Grupo Televisa (media network), Liberty Global (communications company), UPS, Walmart, Alphabet (which owns Google and YouTube), just to name a few.

In January 2021, The Land Report announced that "Bill Gates now owns the most farmland of anyone in the United States. . . . Gates owns land in Washington, California, Idaho, Wyoming, Colorado, New Mexico, Arizona, Nebraska, Iowa, Wisconsin, Illinois, Michigan, Indiana, Ohio, North Carolina, Florida, Mississippi, Arkansas, and Louisiana."

In her 73-page report, *Earth Democracy: Connecting Rights of Mother Earth to Human Rights and Well-Being of All*, Indian scholar/environmental activist Vandana Shiva dismantles Bill Gates's agenda to reach what he calls "net zero" by 2050.[9]

In the new "net zero" world, farmers will not be respected and rewarded as custodians of the land and caregivers, as Annadatas, the providers of our food and health. They will not be paid a fair and just price for growing healthy food through ecological processes, which protect and regenerate the farming systems as a whole. They will be paid for linear extraction of fragments of the ecological functions of the system, which can be tied to the new "net zero" false climate solution based on a fake calculus, fake science allowing continued emissions while taking control over the land of indigenous people and small farmers. "Net zero" is a new strategy to get rid of small farmers, first through "digital farming" and "farming without farmers" and then through the burden of fake carbon accounting. Carbon offsets and the new accounting trick of "net zero" does not mean zero emissions. It means the rich polluters will continue to pollute and also grab the land and resources of those who have not polluted—indigenous people and small farmers.

In an interview with Dr. Joseph Mercola in March of 2021, Vandana Shiva didn't mince words, "Gates ends up wielding enormous control over global agriculture and food production, and there's virtually no evidence to suggest he has good intentions . . . if in the next decade, if we don't protect what has to be protected . . . and take away the sainthood from this criminal, [he] will leave nothing much to be saved."[10]

It's not just Bill Gates's monopolistic tendencies that are troubling, but some of the characters he does business with also conflict with his humanitarian persona.[11]

In October of 2019, the *New York Times* exposed the mysterious relationship between Bill Gates and known sex trafficker Jeffrey Epstein. According to the *New York Times*, "Gates and Epstein met in person for the first time at Epstein's New York home less than two years after Epstein was released from jail in 2009."[12]

At that time, Epstein had been accused of assault by thirty-six women and girls, some as young as fourteen. In a gross miscarriage of justice, Epstein was only convicted of one count of soliciting sex from a minor, and one count of soliciting prostitution.

The *New York Times* also reported that "Gates had flown on Epstein's infamous private jet, the *Lolita Express*, and sometimes even stayed late into the night at Epstein's NY estate."

As Mikki explained, "What you see in *INDOCTORNATION* is barely the tip of the iceberg in terms of revealing the depth of the Gates/Epstein affair. But because we were committed to featuring only facts that could be verified through basic research, unfortunately, some of the shadiest deals the two billionaires were partnered on had to be left out of the movie. We simply weren't willing to make a single claim that we couldn't back up with hard evidence. As we learned through intense research, when you're one of the wealthiest men in the world, and also one of the largest funders in nearly every major digital news and information platform, you have the power to delete history."

CHAPTER EIGHT

Fact-Checking the Fact-Checkers

Education is not the learning of facts,
rather it's the training of the mind to think.

—Albert Einstein

———

"This is a cognitive dissonant moment which is being imprinted in your brain," Dr. Martin told Mikki as the cameras rolled for *PLANDEMIC: INDOCTORNATION.* "Just like 'Remember the Great Depression,' 'Remember 9/11.'"

"We are being conditioned to have the excuse for unbelievable acts of tyranny, which will be justified by 'Remember 2020,'" he continued. "Your loved ones, those that died, those that are infected, they're being used as cannon fodder, which is the ultimate desecration of their honor and integrity."

"This is also a test of humanity to see how much of our liberty we will let go before we finally draw the line under *enough*," David said.

"So what do we do now?" Mikki responded.

Dr. Martin continued, "This is not a time for us to go in a mob frenzy, find the perpetrators, and haul them into the town square and pillory them," he explained. "This is a moment for us to recognize that every decision that is being made today by any of the conspiring parties made perfect sense in each increment when each one of those decisions was made. The sum of those incremental steps, however, has led to devastation, because they lost touch with their fellow humanity.

"But that's an invitation for each one of us to examine how we're living and how not a single decision we make, not one, in any moment is without consequence. This is our moment to reclaim our humanity."

Four and a half hours after the interview began, it was a wrap—and the entire team was shell-shocked. Instead of being just a piece of *PLANDEMIC 2*, Dr. Martin's interview became the foundation of an entirely new project, *PLANDEMIC: INDOCTORNATION*.

"The original intention was to create a trilogy: three thirty-minute episodes," Mikki said. "Halfway into the editing, as online censorship was becoming more and more intense, I thought, 'We may not get to release a third movie.' I made the last-minute call to join Part Two and Part Three to make one feature-length movie." *INDOCTORNATION* is the result.

Due to the impact of Part One, this time around they had the support of hundreds of top doctors and scientists from around the world. Mikki used them all as de facto fact-checkers. "I would write a section of the film, then send the rough script to three different email threads to get feedback," he explained. "One thread was occupied by a dozen or more doctors and vaccine experts. Another was occupied by legal scholars and patent attorneys. The third thread was a mix of researchers and journalists. I asked all three threads to scrutinize every claim until we got it right."

Mikki continued, "Next step was to create a rough edit of the section we were working on, then send that out to the collective to be further scrutinized. Once everyone was satisfied with the information, we would then insert that section into the body of the full movie."

The *PLANDEMIC* production team and their researchers were ready for the challenge. They scoured through archives of obscure medical journals, investigated official government reports, read piles of legislation, studied tax records, retrieved patents, and conducted off-record interviews with eye witnesses. For every major claim, there was a day or a week of work to ensure that it was airtight.

Typically, a feature-length documentary takes one to five years to complete. With *INDOCTORNATION*, they delivered it in a little more than three months. The late nights were worth it if it meant getting the film out during the pandemic, while there was still time to inform the public of the potential hazards of volunteering for a global medical experiment. One of the most basic human rights is the right to *informed consent,* but that right too often gets placed aside when the public is paralyzed by fear.

Heading into the release date, the team was feeling good. They had learned a lot from the lessons of Part One and created what they felt was a strong riposte to the critics. When *PLANDEMIC 1* was released, the attacks on Dr. Mikovits were fast and furious—mainly because she was such low-hanging fruit.

Dr. Mikovits had been dragged through the mud during other scuffles with Big Pharma in the past. Even just one of those older articles, at most outlets, would be enough to justify a new hit piece. It didn't take much critical thought or reporting know-how to follow that line of attack.

Dr. Martin, however, didn't have that kind of baggage. He is a financial analyst, and the founder of the world's first quantitative index of public equities, the IQ 100 CNBC Index on Wall Street. He is a Batten Fellow at the University of Virginia's Darden Graduate School of Business Administration. He served as chair of Economic Innovation for the UN-affiliated Intergovernmental Renewable Energy Organization. He was an advisor to numerous Central Banks, global economic forums, the World Bank and International Finance Corporation, and national governments, including the US Congress.

With that kind of background, the fact that he was stepping out and stepping up was on its own a big deal. In so many ways, Dr. Martin *was* the establishment. He knew the ins and outs, and he knew, in a sense, where the bodies were buried. So, anyone looking for an easy Google search result to pick up and parrot would have a hard time this time around.

The *PLANDEMIC* team knew what to expect from Big Tech. This time they built their own decentralized website, ensuring that it would be far more challenging to take down. In an effort to activate the buzz prior to the official release, they shared the news of *INDOCTORNATION*'s impending launch with the list of followers that had been generated through *PLANDEMIC 1*. They requested that their audience spread the word about where to watch the livestream debut.

Friday, August 18, would be the day for the world premiere of *PLANDEMIC: INDOCTORNATION,* on the popular UK-based platform London Real. In the days before the premiere, London Real's announcements were shared and liked thousands of times across all of the major social media platforms. By announcing the video's release, though, they also alerted the critics—and the trolls came ready, too. As they prepared to start the livestream, London Real's developers sent the *PLANDEMIC* team the following email:

SUBJECT LINE: DOS ATTACK

We are experiencing a TON of brute force denial of service attacks. See the attached log. Someone is trying to disrupt access to our site. We got through it, but it shows the lengths that they are willing to go to stop this from coming out.

London
7:30 p.m. local time, August 18

"Welcome to the world premiere of what might be the most important documentary you will ever see," said Brian Rose, the host and founder of London Real. In the end, Brian's debut of the film went off without a hitch. *PLANDEMIC: INDOCTORNATION* ended up setting the world record for the largest livestream broadcast of a documentary film, with more than 1.9 million unique viewers tuning in for the big event. By the end of the day, the movie was viewed more than four million times on the London Real platform alone. After the livestream finished, the *PLANDEMIC* team released the movie on various platforms. Of course, they knew that just like last time, the hammer of social media censorship would fall at some point. They just had no idea how fast.

Before the movie was even done premiering, Facebook blocked the livestream, and critics were already claiming that the movie had been debunked. Unable to name a single inaccurate claim, critics resorted to lazy one-liners:

NBC News: "Boring"

PolitiFact: "Yawn"

BuzzFeed News: "A Flop"

The only real flop was their plan to prevent people from watching and sharing the movie. By the end of the day, the London Real link was shared more than 300,000 times.

Mikki made the directorial choice to avoid all hot button distractions such as masks, social distancing, or anything related to vaccine safety and efficacy. As a result, this time, the critics would have to find fault with the information itself.

Most reporters were simply too lazy for that, clearly. The few who were foolish enough to attack Dr. Martin's claims often made critical errors such as misreading the patents or misinterpreting key details. Even patent lawyers struggled to understand, and many reached out. Dr. Martin was ready to answer them.

"David began reaching out to all the fact-checkers and critics," Mikki explained. "I was copied on every email. I would wake up every morning and quickly check the thread to see who David was debating that day. He always began engagement with a cordial invitation:

Dear Sir or Madam:

Thank you for covering *PLANDEMIC: INDOCTORNATION.* However, we believe that there seemed to be some misinterpretation of key points. We've attached the original source documents to clarify. Please feel free to reach out should you have any further questions. We would, after review, appreciate a public correction.

Kind regards,
The PLANDEMIC Team

"Only a couple were confident enough to respond," Mikki continued. "Typically, the fact-checker would send a link to a bogus study. David would show them all the ways their evidence was invalid. They would then disappear and stop responding to our messages. Not one had the integrity to offer a retraction, or to even acknowledge their error."

CHAPTER NINE

Ending Where It Began

Without freedom of thought, there can be no such thing as wisdom—and no such thing as public liberty without freedom of speech.

—Benjamin Franklin

A t that moment, America's greatest challenges still lay ahead of her. Over the next six months, the reported COVID-19 death toll would skyrocket. Frustration, anger, and fear ran high across the country, and an unprecedented level of political turmoil brought it all to a boiling point. For some, the January 20 presidential inauguration of Joe Biden and Vice President Kamala Harris marked the beginning of a new era. On his first full day in office, though, President Biden cautioned Americans from succumbing to any undue optimism. Another surge was still under way, and Americans across the country were being asked to hold steady underneath the weight of lockdowns, mask mandates, and the painful surreality of life in a pandemic.

For so much of 2020, the word *vaccine* had been a rallying cry for those in search of a light at the end of the tunnel. President Trump's *Operation Warp Speed* was tasked with developing an effective COVID-19 vaccine in a matter of months—despite the fact that the typical development period for a vaccine is several years.

In December 2020, a vaccine by Massachusetts-based pharmaceutical company Moderna was approved by the FDA for distribution (the first time that the company had *ever* managed to get a product FDA-approved). That was followed closely by a vaccine by Pfizer, and vaccines by Johnson & Johnson and AstraZeneca waited around the corner. Far from a silver bullet, though, the vaccine rollout was plagued by the same inefficiencies, issues, and straight-up errors that characterized so much of the COVID-19 response.

Dr. Martin explained in a January 2020 interview, "The problem is, it's a defense contractor, ATI, who is the official owner of the distribution of the vaccine. People just don't pay attention to the facts that are actually visible right in front of their face, and this is hidden in plain sight. There's nothing secret about what's going on with these guys.

"ATI (Advanced Technology International, Inc.) is a South Carolina company that is directly involved not only in Operation Warp Speed implementation, but—far more important—they are the ones who have the contract for government propaganda and misinformation management," he continued. "You give a company who has a history of providing the Defense Department with misinformation and propaganda management the contract for the vaccine rollout. Is it any surprise that they have no experience, no infrastructure? That's not a shock."

Even if the infrastructure failures had been predicted, there was another bump in the road that wasn't a big surprise to government officials: the frontline workers—those heroes who'd toiled for months and would be rewarded with the first vaccines as a result? Many didn't want it.

A report by Surgo Ventures found that 15 percent of surveyed health-care workers across the country planned to decline their dose. In other cases, the number of refusals was reported as high as 50 percent.

Understandably, there also was widespread mistrust of the vaccine among Black Americans, which led to early imbalances in vaccine distribution. In December of 2020, *Time* reported that of the 350,000 people who had registered online for the vaccine, only 10 percent were Black or Latino. Those two groups account for more than 30 percent of the US population.

Through it all, Dr. Martin was vocal online, urging people to educate themselves on the true nature of the "vaccine." The legal definition of "vaccine" is "a preparation of killed or attenuated living microorganisms, or fraction thereof, that upon administration stimulates immunity that protects against disease."[1]

The COVID-19 vaccine (at least, the leading vaccines by Moderna and Pfizer) does not have any fraction of the COVID-19 virus—neither living nor dead. Instead, it is comprised of mRNA, or "messenger RNA." Messenger RNA is genetic material that normally corresponds to part of your own DNA in cells. It helps with protein synthesis within cells and the body.

The COVID-19 vaccine shoots foreign genetic material, mRNA, into your system. When you receive a COVID-19 shot, that mRNA—instead of creating human proteins, like normal mRNA does—instructs the cell to create a piece of the "spike protein" that is found on the COVID-19 virus.

By definition, that spike protein is "antigenic," or foreign material, so your body attempts to fight it off, thinking it is an infection. The creation of T-cells specific to COVID-19 is intended to prime your immune system for any future viral infection. When the immune response is over, the cell destroys the foreign mRNA with enzymes.

It all sounds pretty harmless, right? The reality is, we don't know yet. This is a new technology that has never been licensed at all in the United States, let alone implemented at mass scale. The long-term effects are unknown.

Former vice president of Pfizer Dr. Michael Yeadon is one of countless whistleblowers who have risked it all to warn the world of the potential dangers. In a Planet Lockdown interview series by James Henry, Yeadon explains,

> You've been subjected to propaganda and lies, by people who are very well trained in how they do that. . . . If you want to check any of the things I've said, you will find it to be true. And I will point out to you, that if you find one thing your government has said which is clearly not true, I ask you this: Why would you believe anything else they've told you?
>
> We're probably quite used to politicians occasionally telling white lies and we kind of let them. But when they lie to you about something technical, something you can check, and they do so repeatedly over months . . . please, you've got to believe me, they're not telling the truth. And if they're not telling the truth that means there's something else afoot.
>
> I'm here today to tell you that there's something very very bad happening. And if you don't pay attention you will soon lose any chance to do anything about it. Don't say you weren't warned.[2]

One of the key inventors of mRNA vaccine technology, and one of the world's foremost experts in vaccines, preclinical discovery research, gene therapy, bio-defense, and immunology, Dr. Robert Malone, sent shockwaves through the scientific community when he took a stand against COVID protocols.

> I don't mean to sound alarmist, but what seems to be rolling out is the worst-case scenario where the vaccine in the waning

phase is causing [the] virus to replicate more efficiently than it would otherwise, which is what we call antibody dependent enhancement. . . . What is antibody dependent enhancement? Briefly, it's that the vaccine causes the virus to become more infectious than would happen in the absence of vaccination. . . . This is the vaccinologists' worst nightmare!

I've been through outbreak after outbreak. I've never seen anything like this. . . . This is behavior control. It's really psyops is what's happening. It's applied psychological operations to control people and their behavior so that they will accept these products, which are still experimental based, on technology that has never been deployed at this level. And as the data are coming out it's becoming more and more clear that these products are not completely safe. . . .

I'm accused now of being an anti-vaxxer and prompting disinformation, but to my eyes the government is obfuscating what's happening here. . . . I'm the opposite of an anti-vaxxer, I'm a true believer. But I'm also committed to safety and good science.[3]

Dr. Martin was insistent when I spoke to him in January 2021: "I have made the point several times: This is not a vaccine. This is a gene therapy that is being marketed under a deceptive medical practice as defined by the Federal Trade Commission. It is being passed off as a vaccine, but it has nothing to do with vaccination under any legal definition of vaccination. Legally, it has to stimulate immunity and stop transmission of a pathogen. Neither of those are being done with this gene therapy."

Why call it a vaccine, then? To Dr. Martin, the choice of words was telling. "If people heard that this was gene therapy, or a form of chemotherapy, then they might think twice," he said. "But if you call it a vaccine and mislead people into thinking that there's a public benefit being served, it's the willful deception of millions of people."

Meanwhile, the obfuscation of the true origin of the virus continued. The WHO first began discussions regarding an agenda for investigation back in February 2020. By July, they announced that the WHO, "together with the Government of China, are setting up an international multidisciplinary team to design, support, and conduct a series of studies that will contribute to origin tracing work."[4]

China was of course in the spotlight, since there was little doubt that the pandemic first sparked there, with Wuhan at the center of it all. This complicated matters. Suspicious of outsiders and viciously dedicated to promoting their version of the truth, the Chinese Communist Party had been accused of widespread lying and misinformation regarding COVID-19 since the beginning of the outbreak. There were allegations that the regime had underreported the size of their outbreak, the contagiousness of the virus, and the nature of its transmission. According to critics, the Chinese had blood on their hands.

Understandably, the Chinese were wary. And understandably, the global community pressed forward with calls for the investigation. By fall 2020, they seemed to have reached an agreement, as the WHO released a specific plan for investigating the roots of COVID-19 in China and beyond. On January 14, investigators began their investigation in China. The Chinese government, however, was determined to temper expectations from the start, insisting that the investigators would merely "exchange views" with their Chinese counterparts—*not* gather evidence.

Although both the Chinese and the WHO continued to ignore the possibility that the virus had escaped from the Wuhan Institute of Virology, the State Department was definitive: "The CCP's deadly obsession with secrecy and control comes at the expense of public health in China and around the world," the report read. "The previously undisclosed information in this fact sheet, combined with open-source reporting, highlights three elements about COVID-19's origin that deserve greater scrutiny."[5]

First, the State Department highlighted "illnesses inside the Wuhan Institute of Virology (WIV)." According to the Fact Sheet, "several researchers inside the WIV became sick in autumn 2019, before the first identified case of the outbreak, with symptoms consistent with both COVID-19 and common seasonal illnesses." WIV senior researcher Shi Zhengli had previously claimed that there had been "zero infection" of any staff or students in the run-up to the pandemic.

Such an accidental infection would not be without precedent, the State Department reminded readers. In fact, there had been several high-profile instances of security lapses at Chinese labs that resulted in infection, "including a 2004 SARS outbreak in Beijing that infected nine people, killing one," the report pointed out.

Second, and crucially, the WIV had been actively focusing on research involving the bat coronavirus since 2016 and up through 2020. They even had a published record of completing the kind of "gain-of-function" research that can help such viruses make the jump to human infection.

Third was the factor that many nations were too scared to even consider: China's potential for biological warfare. "For many years, the United States has publicly raised concerns about China's past biological weapons work, which Beijing has neither documented nor demonstrably eliminated, despite its clear obligations under the Biological Weapons Convention," the State Department report read.

"Despite the WIV presenting itself as a civilian institution, the United States has determined that the WIV has collaborated on publications and secret projects with China's military. The WIV has engaged in classified research, including laboratory animal experiments, on behalf of the Chinese military since at least 2017." A heavy allegation, the State Department said it was supported by the evidence that they had gleaned despite the country's history of "secrecy and nondisclosure"—and it was only scratching the surface.

The report concluded with a pledge to continue to demand transparency from the Chinese, along with a host of other proclamations.

First, noting that "the CCP has prevented independent journalists, investigators, and global health authorities from interviewing researchers at the WIV, including those who were ill in the fall of 2019," the State Department insisted, "Any credible inquiry into the origin of the virus must include interviews with these researchers and a full accounting of their previously unreported illnesses."

Second, "WHO investigators must have access to the records of the WIV's work on bat and other coronaviruses before the COVID-19 outbreak. As part of a thorough inquiry, they must have a full accounting of why the WIV altered and then removed online records of its work with RaTG13 and other viruses."

And third, "The United States and other donors who funded or collaborated on civilian research at the WIV have a right and obligation to determine whether any of our research funding was diverted to secret Chinese military projects at the WIV."

The State Department report concluded with a promise: "As the world continues to battle this pandemic—and as WHO investigators begin their work, after more than a year of delays—the virus's origin remains uncertain. The United States will continue to do everything it can to support a credible and thorough investigation, including by continuing to demand transparency on the part of Chinese authorities."

Less than a week after it was published, it was gone from the State Department website. Apparently, the Chinese aren't the only ones with a proclivity for what could kindly be called "information management."

To Dr. Martin, saving lives by ending this pandemic and saving our democracy depends upon transparency and the sharing of information. "I have made sure to put published, accessible information into the public view," he said in January 2021.

"That is seen as this massive public service that I'm doing. The truth is, it's not much of a public service. It's what accountability of a citizen should be. We should pay attention to these things. One of the facts that I have been quite insistent on is that we live in an era where we've abdicated the responsibilities of living in a democratic society," he continued.

"Instead of reading source documents, we're checking social media. 'Following' people has led us down a pathway of checking our own intellect and power of inquiry at the door. We're assuming and trusting that someone or something else has inside knowledge.

"I would say my entire passion, and what got me involved in *PLANDEMIC* in the first place, was about making it very clear to the public that their own powers of inquiry are a muscle that need to be engaged," he continued.

"The Spanish are unique in that their Constitution states that literacy is a prerequisite for democracy. 'Literacy' does not mean the very capacity to read a Twitter feed, but to formulate an independent question, to create an independent hypothesis, and to go out and find the information to test that." He concluded, "Our democracy will only survive if we cultivate that kind of literacy."

I reached out to Dr. Mikovits to request a closing statement. This is what she had to say:

> I gave a talk in September of 2018, in Phoenix. At the end they asked, "Judy, is there anything else you want to say?" I'm looking at the pained expressions on the faces of moms and dads in the audience, sitting there with their injured children.
>
> I said, "I was part of the problem. I was part of the system that is responsible for hurting million of innocent people. Innocent children." And I paid the price. Yes, I was targeted and terrorized for refusing to play on the darkside. Yes, they took my home and my life's savings. Yes, I had my name and reputation ruined. And yes, this has been the best decade in my life!

I thought I belonged hidden away in a laboratory. I was a science nerd. I never imagined that someday I would talk for a living. That my story would be my medicine.

If you just stand up and speak the truth in love you'll be honored. I've been honored in so many ways, I can't even begin to describe it. We can win this thing. I'm living proof of that. Courage is contagious.

If you're a doctor who is just waking up to the awareness that you've violated your oath to "First do no harm," or if you're a parent of a child that through your guidance has been damaged by vaccines, the highest choice you can make in this moment, as painful as it might be, is to acknowledge your mistake and forgive yourself.

You listened to the science. You did everything you were taught and told to do. Take responsibility, but let go of shame. It belongs to someone else. Forgive yourself like I forgave myself.

Epilogue

If we surrendered to earth's intelligence we could rise up rooted, like trees.

—Rainer Maria Rilke

In the months since beginning this book, a peacefully unified country seems like even more of a distant dream. Contrary to our hopes (and I say that as someone who voted for Joe Biden), the end of the Trump presidency did little to quell the waves of division sweeping through our nation. If anything, the results of the 2020 election only incited even more.

As many on the Republican side questioned the results of the election, Trump supporters looked to the Congressional counting of the electoral college votes on January 6 as the last chance for a Trump upset. Thousands of supporters arrived in Washington, DC, on that date to make their voices heard.

On the surface it was intended to be a demonstration of the will of the people, a last stand for the MAGA crowd. Like so many, I was disturbed to see it all unfold. I also was shocked to see Mikki at the center of it.

The *New York Times* posted a photo of Mikki in the heart of the melee, claiming he "joined the siege on the Capitol." They also called him "a video producer who made a popular video filled with falsehoods about the coronavirus." Knowing as I did that at least that part of their depiction was untrue, I was curious to get the real story behind Mikki's participation in such a terrible incident. At least, I was willing to give him the benefit of the doubt and to ask him to explain.

As I was wrapping up my work on this book, Mikki called me. Chastened by having been labeled an insurrectionist—although, ultimately, not too surprised—he was relieved and hopeful to share for the first time what really happened and why he was there that day.

Mikki began, "I was invited to speak at an event called Health Freedom." Already planning to be in D.C. to film interviews for a project about the effects of the pandemic lockdowns, Mikki found the invitation serendipitous. The event organizers sent Mikki a flyer featuring the lineup of speakers. Impressed to see several doctors and health experts whom he respected, he accepted the invitation.

The day prior to his departure, Mikki received an updated event flyer. To his surprise, the acronym MAGA had been added across the top. In addition to that, one of President Trump's former advisers, Roger Stone, had been added as a speaker. Though Mikki knew very little about Roger Stone, he knew enough to understand that the combination of his presence and the MAGA brand could hinder his ability to build bridges between political parties.

Mikki told me, "Having been born and raised on the left, I'm able to connect with people who identify as Liberal but are struggling to align with many of the new ideologies of their party. And, as someone who stands strong for family and freedom, I also connect with many people who identify as Conservative. There is no single box for me. In that sense, I'm a political orphan. And I know I'm not alone."

Uneasy with the last-minute changes, Mikki called the event organizers to express his concerns. "Having spoken at one other Health Freedom event, I knew the organizers to be good people," Mikki said. He was told that the word MAGA was added to get the attention of those attending the MAGA March in hope that they would attend the Health Freedom event. Mikki then expressed concern that Roger Stone's presence during such a volatile moment in time might create issues for the speakers who were only there for a cause as universally important as Health Freedom.

It wasn't just Mikki who spoke up. His good friend, Del Bigtree, host of *The Highwire*, a popular online show that focuses on vaccine awareness, was also concerned. Mikki shared with the organizers, "I just spoke with Del, and he too feels less than comfortable with these last-minute changes."

Without hesitation, the organizers promised they'd halt distribution of the MAGA flyer and assured Mikki that the stage would only reflect branding aligned with Health Freedom. Through another source, Mikki was informed that Roger Stone would not be attending the event. Under these circumstances, with flights and hotel rooms already paid for, Mikki and Del agreed to follow through.

Mikki recounts his entire experience:

"One critical detail the media has kept hidden from the world is the fact that not everyone who traveled to D.C. on January 6 was there for the 'stop the steal' effort. Many were there to peacefully protest the extended lockdowns, vaccine mandates, and the loss of civil liberties. That's why I was there. No other reason. Hence my absence at the rally and the march.

"That morning, as I prepared for my talk in my hotel room, I directed my two-person crew, videographers Keresey and Sarah, to go out and get interviews with people they wouldn't expect to see there. I wanted to hear what immigrants had to say. I wanted to hear what people of color had to say. Knowing that the media would only focus on the bad news and bad people, I wanted to capture uplifting stories from good people, particularly those who immigrated to America.

'My goal was to create a short film that would remind Americans just how fortunate we all are to live in such a diverse and progressive nation. Around lunch time, I made my way over to the stage. Sarah and Keresey were already there with cameras rolling.

'I was connecting with people backstage when someone whispered in my ear, 'The Capitol is being stormed.' The stage was just a few blocks from the Capitol building. I told Keresey and Sarah to head over to see if the rumor was true. They hustled off. About thirty minutes later, I began to hear sirens in the distance. I attempted to radio Keresey and Sarah, but they didn't answer. The sirens intensified. I was getting worried. I asked the event organizers if I could speak at a later time so I could go check on my crew. I jogged to the Capitol, expecting to see mayhem.

"Having never been to the Capitol, I was unaware that I was on the back side. The mayhem I was expecting had already occurred on the front side. The back side was a very different story. Everyone was smiling, waving flags, and taking selfies. There were families with babies on shoulders and in strollers. The crowd was far more diverse than even I was expecting. Once again, I tried to reach my crew by radio, but they didn't respond.

"Police were holding the crowd behind barricades, not far from the staircase that leads to the back doors of the Capitol building. Police and citizens were communicating on the front line. I moved closer to hear what they were saying. Various people were pleading cordially with law enforcement, saying things such as, 'The lockdowns are killing us. . . . We're going to lose our family business. . . . I can't feed my children. . . . We're not your enemy.'

"One guy was repeating over and over, 'We love you. We love you. We have nothing but love for you.' Another man was passionate, but calm, when he said, 'Why is it that we're being held here, while every other protest group gets to stand on the stairs of the Capitol to have their voices heard? That's not right. We pay our taxes. We deserve respect, too.'

"Suddenly, as if receiving the command over their earpieces, in unison, police opened the barricades, allowing the crowd onto the stairs. I was moved. To see people use their voices in such a conscious manner and to witness it producing a peaceful and positive result was inspiring. Just as I stepped onto the stairs, my cell vibrated. It was one of the event producers telling me to return quickly, as I was up next.

"I rushed back, just in time for my talk. Still moved by what I had witnessed, I opened with: 'I'm a little out of breath because I was just a part of this situation. Our proud patriots just pushed past riot police, peacefully, as peacefully as that could happen.'

"In retrospect, I regret using the word 'pushed.' To be clear, I saw no physical pushing whatsoever. I used that word in the way one might say they *pushed* through a cold or a long day. The other word I used that stirred controversy was 'patriots.' That word I do not regret using. A patriot is simply a citizen who loves their country. I love my country and I will not allow anyone to take that away from me.

"Having traveled the globe, I've seen the best and the worst this world has to offer. Yes, America has a horrific past and still has many issues that need our immediate attention. But does anyone honestly believe we can fix our current issues through shame and hate? It's those very energies that got us here. Seeing that is the first step toward healing the wounds of our past.

"While I was onstage the sound of distant sirens grew even louder. I could barely focus on my talk. The moment I was done, I bolted, returning to the back side of the Capitol. I was relieved to see that everyone was still in cheerful spirits.

"I asked an elderly Asian woman what was happening. She said with a big smile, 'They're letting people in now.' I thanked her, then made my way to the stairs. Halfway up, I stopped to scan the crowd from this higher vantage point. Keresey and Sarah were nowhere in sight.

"I gave my radio one last try. Still no reply. I approached the back doors of the Capitol. Sure enough, police were allowing people to filter in and out. In this location emotions were mixed. People were both smiling and frowning. An older woman was crying.

"Like before, police were communicating with people on the frontline. I moved closer to hear what they were saying. That's when I saw that the windows within the doors were cracked. This was the first indication that some level of force had been used.

"The negotiations in this location were a mix of peaceful and instigative. The same guy from down below was there and still repeating, 'We love you. We love you.' Another guy toward the back was yelling, 'Time's up! Get out!' It wasn't clear if he was speaking to the police or to the meandering crowd that was blocking others from entering the building. As much as I wanted to see if my crew was in there, under the circumstances my intuition told me to stay put. I began filming the moment with my cell phone.

"Two guys began pushing the crowd from behind. I yelled loudly, 'Hey hey hey hey, whoa, don't do that shit! No no no No!' They stopped. Just as things were settling, the crowd began to chant something. At that same moment, my earpiece was filled with radio chatter, making it difficult to hear. It wasn't until later that I realized they were chanting, 'Hang Mike Pence.'

"Luckily, there was a camera on me that whole time, providing proof that I took no part in that chant. As a veteran activist, I've seen what works and what doesn't. Chants of this nature, or any form of radicalism, is counterproductive at best.

"Once again, the same two guys began pushing the crowd from behind. This time police responded with pepper spray. I was hit in the eyes and mouth. Unable to see, I felt my way out of the crowd, then sat down on the stairs. Kind people helped wash my eyes.

"Once I could see again, I stood up to record a video diary. This is something I've learned to do in situations like that. Oftentimes when we're involved in something so chaotic, it becomes difficult to remember the succession of events, which makes editing an accurate timeline all the more challenging.

"As I was recording that video, I noticed a guy with a video camera covertly filming me. He didn't appear to be a professional, so I didn't think much about it. As it turned out, he was a *New York Times* videographer apparently assigned to stalk me.

"He filmed me making that video diary entry, which was ultimately used out of context to make it appear as if I were in favor of the violence. I wouldn't know this, though, until a couple of days later.

"I was still on the stairs and recovering from the pepper spray when my phone rang. It was Sarah and Keresey. They were safe and waiting for me back at the stage. Relieved, I hustled back to reunite with them. Together, we returned to our hotel. The lobby was filled with people. Emotions were heavy. That was the first moment that we learned about the violence that took place on the front side of the Capitol, and that a woman had been killed. Like everyone else in that lobby, we were devastated.

"We returned home the next day and immediately began editing our footage. It wasn't until January 8 that I became aware that certain media outlets were fabricating a narrative to further smear my name. This is what happens when you dare to expose corporate propaganda empires.

"As a veteran media producer, I know their game, so I'm never surprised at how low they will stoop for clicks, ratings, and political leverage. In *PLANDEMIC: INDOCTORNATION*, I pulled back the curtain to expose their dirty and divisive game. January 6 was their attempt at payback."

***

Mikki was contacted by the *New York Times* for an interview about his role on that day. Wary of their intentions, he agreed to a written interview only—one that they still managed to spin. Here's an excerpt from that exchange:

NYT: Did you actually get inside the Capitol? Can you describe what you did there?

MIKKI: No, I did NOT go inside, though I could have. Police were allowing large groups to enter through various doors.

Despite the clarity, this is what made it to print on January 12, 2021, in the *New York Times*: "Mr. Willis entered the Capitol building, but said in a Facebook post that he did not go in far and left quickly."[1]

Mikki told me, "Never have I said any such thing in a Facebook post. The US Capitol is one of the most heavily surveilled buildings in the country. If I had gone inside, that would've been caught on camera. That blatant lie put my freedom and life in danger."

He continued, "Dishonest publications like the *New York Times* don't care if their lies are obvious. They know that the vast majority of readers will believe whatever they print without inquiry. Oftentimes, even after the truth reveals the opposite reality, most people will stick to their initial judgment. That one dishonest post created an avalanche of attacks on me.

"Online trolls grabbed still images of me during the moment the crowd was chanting, 'Hang Mike Pence,' then added the caption, 'Meet Mikki Willis, creator of PLANDEMIC, and domestic terrorist,'" he said. "Another meme declared that I was the one leading that chant. Naturally, this further riled the already hysterical mobs. It didn't end there. A shady social media channel edited the video of my January 6 talk, conveniently cutting out the part where I equally criticized both political parties to make it appear as if I were only critical of the left."

He added, "These are a few of the words that were omitted from that clip: 'This is far beyond a partisan issue and that's what I came here to speak to you about today—the necessity for us to see beyond the mind control that has us still believe in this left-right-blue-red BS.'"

After a deep breath, Mikki settled back and expressed his thoughts on the aftermath.

"While the attacks have weighed heavily on me at times, I'm not angry. Not at the people, that is. Certain media outlets have earned my disdain, but in terms of the public, I hold no resentment. In a strange sort of way, absorbing the blows has left me stronger and more certain about our future.

"When haters come at me, I do my best to remember that at the core of their rage is the primal impulse to live. If *PLANDEMIC* was as dangerous as the critics claimed it was, people responded exactly the way they should have.

"That's why, when I see mobs of pitchforks headed my way, I get excited. I get excited because I want to see the people rise for the honor of life. I want to see the masses use their immense power to stand for the lives of their loved ones. I get excited because I know it's only a matter of time before the truth reaches critical mass and the people redirect their power to the actual dangerous ones.

"Yes, there are dangerous people among us. While they are few in numbers, over the course of generations they've accumulated great power over our lives. Through endless mergers and acquisitions, they now own and control the vast majority of the media, the entertainment industry, the medical industry, big tech, education, our food supply, our energy systems, sports, politics, and so much more.

"It's critical that we are aware of this, but that we don't allow fear to consume us. When the curtain was pulled back for me a few years ago, I went through a process that can only be compared to 'The Five Stages of Grief.' Having lost loved ones, I'd been there before.

"Those stages, as originally mapped by Swiss psychiatrist Elisabeth Kübler-Ross, are Denial. Anger. Bargaining. Depression. And finally, Acceptance. In my experience, they don't always appear in that order, but at some level, everyone in their own way passes through each of those emotions when processing deep grief.

"There's a lot to grieve in this moment. Our reality has been shattered. Our world has been inverted. Up is down. Good is bad. Light is dark. Many of the people we believed to be heroes are being revealed as villains. That's never an easy thing to wake up to.

"The good news is, we're waking up. The human organism is awakening like never before. We've been asleep for generations. It may take us a moment to stand up. But when we do, we will be at a new and higher altitude to see farther than we've ever been able to see before.

"I know this because as a storyteller I've studied the history of human mythologies. There's a lot we can learn from the stories we've been sharing for thousands of years. That is, the *story* we've been sharing. One story with infinite variations. Whether it's a classic like *The Wizard of Oz*, or the latest *Marvel* adventure, the moral of the story is almost always the same: *Follow the yellow brick road (your intuition). You are the one. The force is within.*

"Why is it that our most iconic allegories almost always remind us to look inward for salvation? Is it possible that our wise ancestors knew that a day would come when humanity would be severed from its own Nature? Personally, I believe this is one of our deepest collective wounds. Dig at the root of every bad apple, and you'll discover an agenda to move us farther and farther away from the life-sustaining brilliance of Nature. Some might call that God.

"Remember, before John D. Rockefeller, we had only one pharmacy—Mother Earth. He knew that by breaking the connection between us and our planet, he could create life-long customers. Rockefeller gave birth to a toxic trend that continues today.

"One of the most frequently asked COVID-related questions is 'But why would they destroy their own economy, kill jobs and small businesses?' Short answer: Dependency. It is all about dependency. Soon, should we continue to allow it, most citizens will be dependent on a monthly stipend. After that, say something online that your government doesn't approve of, AND you and your family don't eat that month.

"This is how a microminority assumes total control over a macro-majority. The more we allow the federal government to 'care for us,' the more we lose our civil liberties and sovereignty. We've already reached a point where simply talking about this is dangerous.

"Even more disturbing than the fact that our right to free speech is under attack is that our language is being weaponized. God, liberty, freedom, love, marriage, patriot, man, woman, father, mother, America are just a few of the words currently being burned at the stake. The reason is obvious: these are the things worth fighting for. Remove these pillars of life, and our world comes crashing down.

"That's what they want. Total destruction. After that, they will 'build back better,' as the slogan goes. While that might sound appealing to the uninformed, in reality, their goal is the creation of a One World Government that owns and controls everything—including you.

"In the words of Klaus Schwab, the reigning Führer of The World Economic Forum, 'You will own nothing, and you will be happy.' Half of that statement is absolutely true. Those who perceive themselves to be on the 'good side' believe they'll be spared and invited to live among the chosen in their sparkly new utopia. History offers a much bleaker outcome for 'useful idiots,' as Soviet dictator Vladimir Lenin labeled them.

"The murderous end of the Roman Senators, King Philip IV's extermination of the Knights Templar, Hitler's Brownshirts, and Hugo Chavez's vicious elimination of the very forces he used to put himself into power all show a pattern of the enablers of tyranny being executed by the very swords their socialist leaders convinced them to sharpen.

"Every citizen, US and beyond, must quickly realize that all these new liberty-reducing policies being rushed into law under the guise of public safety will ultimately be used to justify total control over *all* people.

"Only by pitting the people against each other can politicians manipulate a majority into voting against their own freedoms. When 'the left' cheers for censoring 'the right' (or vice versa), they are unwittingly inviting the same punishment upon themselves. When it comes to political takeovers, like the one we're currently experiencing, there is no safe side.

"Today's political playing field is really more of an octagon. A battle cage with multiple sides. In each corner we have Democrats, Republicans, Liberals, Conservatives, Progressives, Libertarians, left, right, far left, far right, and other factions. All of this is by design. The more we're fragmented, the more easily we're divided. The more we're divided, the more easily we're controlled."

Mikki paused and added, "On that note, I want to share something my buddy Cal said the other day that stuck with me. He said, 'Watching your life unfold this year has changed me.' I asked, 'How?' He paused, then looked me in the eyes and said, 'Because I know you. I know your heart. You're one of the most caring and compassionate men I've ever known. It made me see that if they can spin such a hateful narrative against someone like you, none of us are safe.'"

***

Indeed, that last statement from Cal is why you've made it to the end of this book with hopefully very few clues as to my identity. I'm not willing to put myself out there as a prop for someone else's story, and it's not just the media that—I know from experience— is far from impartial. It's the people I don't know, *and* those that I do. Everyone with a social media account has the potential to change someone's life—most often, for the worst. I can't afford to be cancelled.

Does that mean I'm a coward? Now, more than ever, we should be strong enough to speak out against injustice, to hold people accountable with the power of the people. From where I write—and the *PLANDEMIC* team would agree with me—we could all stand to do a little bit more listening than talking. I mean, really *listening*, not just hearing. The world has gotten so loud, our minds so crowded, that we tend to confuse the latter with the former. We hear what newscasters say each night or visually clock the memes we scroll by on our feed, but we never slow down enough to digest the firehose of information that's drowning us. A lot of powerful people would prefer that we didn't.

So, why write a book at all? Maybe it's old-fashioned, but I think that books are one of the last ways in which we truly listen to the voice and viewpoint of another human being. To devote yourself to a one-way conversation for the days and weeks it takes to finish an author's work takes a level of respect, patience, and goodwill that's often absent in our daily conversations—and certainly absent in most Internet interactions. I'm grateful that you've opened your mind enough to consider everything I've laid out in these pages.

This book represents everything that I believe to be true and important about the making of *PLANDEMIC* and the cultural moment that surrounded it. Of course, we're all victims of our own unconscious biases, but coming into this as a skeptic and ending up as a sympathizer has hopefully left me somewhere in the middle, as close to impartial as I can get.

Where is that? I'm definitely not getting the COVID vaccine, but I do wear a mask. I voted for Joe Biden, but I just voted for a Republican in my local race. I'm not a conservative, nor a liberal. I'm a ball of contradictions, just like you.

Humans are infinitely complex, and our opinions can be, too. We don't have to opt for one of the two package deals that seem to be on offer right now: left, or right. If we truly listen to one another and pause before rushing into judgment, we can shape our own opinions, make our own decisions, and prevent others from doing that for us. Perhaps that's the first step toward reclaiming our humanity.

Mikki put it best in the final moments of *PLANDEMIC: INDOCTORNATION*:

> Our lives are shaped and guided by stories.
>
> The stories we're told become the stories we tell.
>
> The more we hear them, the more we believe them.
>
> When used as a tool, they help us to better understand who we are, where we came from, and where we're going.
>
> When used as a weapon, they can be deadly.
>
> One of the most dangerous stories we've been told is the one that goes something like this:
>
> Humanity is a failed experiment.
>
> We are parasites. A cancer. A virus.
>
> It is a myth that permeates our movies, our music, our media, and our minds.
>
> As they say, repeat a lie often enough, it becomes truth.
>
> Fear shuts down the part of our brain designed to solve problems.
>
> Without that ability, we look for others to guide and save us.
>
> In doing so, we lose touch with our most primal Nature.
>
> We forget that we are an extension of the most brilliant and resilient ecosystem in the universe.
>
> We stop eating food grown from the earth and begin consuming products processed from machines.
>
> We trade medicines that heal for drugs that harm.
>
> We abandon love and liberty for debt and dependency.
>
> The good news is, our story is not over.
>
> The climax has yet to come.

That moment when the hero rises from defeat, summoning a force they forgot they had.

A force within.

A force of *Nature*.

They tried to bury us, but they didn't know we were seeds.

—Dinos Christianopoulos

Acknowledgments

To my mother Jackie and my father John. Thank you for the love. Thank you for my life. And to all the brave whistleblowers, scientists, doctors, frontline workers, journalists, elected officials, civil servants, mothers, fathers, and all the everyday heroes who are standing up and speaking out in the name of life, liberty, and the love of humanity, thank you! History will smile upon you.

—Mikki Willis, Father/Filmmaker

Endnotes

Front Matter

1 Kit Stolz, "Mr. Willis Goes to Washington," *Ojai Magazine*,
   Spring 20201, www.facebook.com/ojaivalleynews/posts/mr-willis
   -goes-to-washington-is-a-feature-story-by-kit-stolz-about-former
   -ojai-r/3935269049891583/.

Chapter One

1 Andrew Joseph, https://www.statnews.com/2020/01/11/first-death
   -from-wuhan-pneumonia-outbreak-reported-as-scientists-release-dna
   -sequence-of-virus/.

2 Dr. Joseph Mercola, https://www.organicconsumers.org/news/niaid
   -moderna-had-covid-vaccine-candidate-december-2019.

Chapter Two

1 Jon Rappoport, "Here's what Sharyl Attkisson told me about the 2009 'pandemic,'" April 16, 2020, https://www.eastonspectator.com /2020/04/16/heres-what-sharyl-attkisson-told-me-about-the-2009 -pandemic-apr-16-by-jon-rappoport/.

2 Lyn Redwood, Mary Holland, "Dr. Fauci and COVID-19 Priorities: Therapeutics Now or Vaccines Later?" March 27, 2020, https://childrenshealthdefense.org/news/dr-fauci-and-covid-19-priorities -therapeutics-now-or-vaccines-later/.

3 "AIDS and the AZT Scandal: SPIN's 1989 Feature, 'Sins of Omission,'" Spin, October 5, 2015, https://www.spin.com/featured/aids-and-the -azt-scandal-spin-1989-feature-sins-of-omission/.

4 Bruce Nussbaum, Good Intentions: *How Big Business And the Medical Establishment Are Corrupting the Fight Against AIDS* (New York: Atlantic Monthly Press, 1990).

5 The Village Voice - 1989 - By Larry Kramer - "An Open Letter to Dr. Anthony Fauci" - https://www.villagevoice.com/2020/05/28 /an-open-letter-to-dr-anthony-fauci/.

6 Celia Farber, "Sins of Omission," Spin, November 1989.

7 Bob Herman, "Pfizer raises estimate of COVID-19 vaccine sales by 29%," July 28, 2021, https://www.axios.com/pfizer-covid-19 -vaccine-sales-second-quarter-2021-7bd7ae91-0b1b-4432-be40 -9c91c3dad0dc.html.

8 ABC World News Tonight, television broadcast, Tom Llamas, May 23, 2020.

9 Elaine Cobbe, "France bans use of hydroxychloroquine, drug touted by Trump, in coronavirus patients," CBS News, May 27, 2020, https://www.cbsnews.com/news/france-bans-use-of-hydroxychloroquine -drug-touted-by-trump-to-treat-coronavirus/.

10 Morning Joe, May 22, 2020, https://www.facebook.com/msnbc/posts /hydroxychloroquine-the-antimalarial-drug-touted-by-president-trump -is-linked-to-/3761296230633268/.

11 Mandeep R. Mehra, Sapan S. Desai, Frank Ruschitzka, Amit N. Patel, "Hydroxychloroquine or chloroquine with or without a macrolide for treatment of COVID-19: a multinational registry analysis," *The Lancet*, May 22, 2020, https://www.thelancet.com/journals/lancet/article /PIIS0140-6736(20)31180-6/fulltext.

12 Catherine Offord, "The Surgisphere Scandal: What Went Wrong?" *The Scientist*, October 1, 2020, https://www.the-scientist.com/features /the-surgisphere-scandal-what-went-wrong--67955.

13 Charles Piller and Kelly Servick, "Two elite medical journals retract coronavirus papers over data integrity questions," *Science*, June 4, 2020, https://www.sciencemag.org/news/2020/06/two-elite-medical-journals -retract-coronavirus-papers-over-data-integrity-questions.

14 Catherine Offord, "The Surgisphere Scandal: What Went Wrong?"

15 "Updated: Lancet Published a Fraudulent Study: Editor Calls it 'Department of Error,'" June 2, 2020, https://ahrp.org /the-lancet-published-a-fraudulent-study-editor-calls-it-department -of-error/.

16 Yahoo Sports, "Renowned epidemiologist sees 'massive disinformation campaign' against hydroxychloroquine," August 23, 2020, https://sports.yahoo.com/renowned-epidemiologist-sees-massive -disinformation-005033779.html.

17 "C.D.C. Internal Report Calls Delta Variant as Contagious as Chickenpox," August 1, 2020, https://www.nytimes.com/2021/07/30 /health/covid-cdc-delta-masks.html.

18 "Kary Mullis [PCR Inventor] - The Full Interview by Gary Null [HIV/AIDS]," May 1996, https://www.bitchute.com/video /AHJwHmTiGsOw/.

19 Andreas Stang, MD, MPH; Johannes Robers, MTA, et al., "The performance of the SARS-CoV-2 RT-PCR test as a tool for detecting SARS-CoV-2 infection in the population," https://www.ncbi.nlm.nih .gov/pmc/articles/PMC8166461/.

20 Alex Ralph, "Bill Gates and George Soros buy out UK Covid test company Mologic," *The Times*, July 20, 2021, https://www.thetimes .co.uk/article/bill-gates-and-george-soros-buy-out-uk-covid-test -company-mologic-70c3r736b.

21 January 22, 2005 - "Royalty payments to staff researchers cause new NIH troubles" - By: Janice Hopkins Tanne - https://www.ncbi.nlm .nih.gov/pmc/articles/PMC545012/.

22 https://www.gavi.org/our-alliance/about.

23 "Transcript: Bill Gates Speaks to the FT About the Coronavirus Crisis," April 9, 2020. https://www.ft.com/content/13ddacc4-0ae4-4be1 -95c5-1a32ab15956a.

24 Becky Quick, "Bill Gates and the return on investment in vaccinations," January 23, 2020, https://www.cnbc.com/video/2019/01/23/bill-gates -and-the-return-on-investment-in-vaccinations-davos.html.

Chapter Three

1 Josh Rottenberg and Stacy Perman, "Meet the Ojai dad who made the most notorious piece of coronavirus disinformation yet," May 13, 2020, https://www.latimes.com/entertainment-arts/movies/story/2020-05-13 /plandemic-coronavirus-documentary-director-mikki-willis-mikovits.

2 Sheera Frenkel, Ben Decker, Davey Alba, "How the 'Plandemic' Movie and Its Falsehoods Spread Widely Online," *New York Times*, May 20, 2020, https://www.nytimes.com/2020/05/20/technology/plandemic -movie-youtube-facebook-coronavirus.html.

3 Yaneer Bar-Yam, "Don't rebreath the coronavirus: New mask designs," New England Complex Systems Institute, April 3, 2020, https://necsi .edu/dont-rebreath-the-coronavirus-new-mask-designs.

Chapter Four

1 Paul Elias, "Race to patent SARS virus stirs debate, *Associated Press*, May 5, 2003.

2 Ibid.

3 Ibid.

4 "Scientists race to patent SARS virus," *Associated Press,* December 15, 2003, https://www.nbcnews.com/id/wbna3076748.

5 Isabel Vincent, "COVID-19 first appeared in a group of Chinese miners in 2012, scientists say," August 15, 2020, https://nypost.com/2020/08/15 /covid-19-first-appeared-in-chinese-miners-in-2012-scientists/.

6 "Statement on funding pause on certain types of gain-of-function research," October 16, 2014, https://www.nih.gov/about-nih/who-we -are/nih-director/statements/statement-funding-pause-certain-types-gain -function-research

7 Fred Guterl, "Dr. Fauci Backed Controversial Wuhan Lab with U.S. Dollars for Risky Coronavirus Research," *Newsweek*, April 28, 2020.

8 Bob Roeher, "WHO wades into row over sharing of H5N1 flu research," *The BMJ*, January 4, 2012.

9 Marc Lipsitch, "The U.S. is funding dangerous experiments it doesn't want you to know about," *Washington Post*, February 27, 2019.

Chapter Five

1 March 2015 - National Academies of Science, Engineering & Medicine - https://www.nationalacademies.org/our-work/enabling-rapid-medical -countermeasure-research-discovery-and-translation-for-emerging -threats-a-workshop.

2 https://www.judiciary.senate.gov/download/epstein-testimony.

3 Ibid.

4 Richard Epstein, "Google's Hypocrisy," Huffington Post, October 6, 2015.

5 Molly Ball, "The Secret History of the Shadow Campaign That Saved the 2020 Election," *TIME*, February 4, 2021, https://time.com/5936036 /secret-2020-election-campaign/.

6 Steve Lohr, "Data engineer in Google case is identified," *New York Times,* April 30, 2012.

7 Paul Roberts, "Tacoma-based Snopes, debunker of fake news, is locked in nasty legal battle," June 4, 2009, https://www.seattletimes.com /business/tacoma-based-snopes-debunker-of-fake-news-is-locked-in-a -nasty-legal-dispute/.

8 Kate Murphy, "Single-Payer & Interlocking Directorates," July 2009, https://fair.org/home/single-payer-and-interlocking-directorates/.

9 Carl Bernstein, "The CIA and the Media," *Rolling Stone*, October 20, 1977, http://www.carlbernstein.com/magazine_cia_and_media.php.

10 Chase Peterson-Withorn, April 2021 https://www.forbes.com/sites /chasewithorn/2021/04/30/american-billionaires-have-gotten-12-trillion -richer-during-the-pandemic/?sh=59fc02e6f557.

Chapter Six

1 Widely available video clip of the proceedings: https://www .centerforhealthsecurity.org/event201/videos.html.

2 Center for Health Security - "Event 201 Pandemic Exercise Underscores Immediate Need for Global Public-Private Cooperation to Mitigate Severe Economic and Societal Impacts of Pandemics," October 17, 2019, https://www.centerforhealthsecurity.org /event201/about.

3 Norbert Häring, "Why is Gates denying Event 201?" *National Herald*, May 2, 2020, https://www.nationalheraldindia.com/international/why -is-gates-denying-event-201.

4 Katie Engleman, "Sherlock Biosciences Receives FDA Emergency Use Authorization for CRISPR SARS-CoV-2 Rapid Diagnostic quad-shape COVID-19 Test is First FDA-Authorized Use of CRISPR Technology," Sherlock Biosciences, May 7, 2020, https://sherlock.bio/sherlock -biosciences-receives-fda-emergency-use-authorization-for-crispr-sars -cov-2-rapid-diagnostic/.

5 Ed Cara, "How last year's pandemic simulation foreshadowed Covid-19," Gizmodo, October 26, 2020.

6 Zosimo T. Literatus, "'Plandemic' Fact Check: U.S. Patent on Coronavirus," Yahoo News, March 2, 2021, https://ph.news.yahoo.com/literatus-plandemic-fact-check-u-140100264.html.

7 Zosimo T. Literatus, "'Pandemic' Fact Check: Conclusion," Yahoo News, June 1, 2021, https://ph.news.yahoo.com/literatus-pandemic-fact-check-conclusion-110200556.html.

8 "Event 201 Pandemic Exercise Underscores Immediate Need for Global Public-Private Cooperation to Mitigate Severe Economic and Societal Impacts of Pandemics," Center For Health Security, October 17, 2019, https://www.centerforhealthsecurity.org/event201/about.

9 Gerard Gallagher, "Fauci: 'No doubt' Trump will face surprise infectious disease outbreak," *Infectious Disease News*, January 11, 2017.

10 Nicola Twilley, "The terrifying lessons of a pandemic simulation," *The New Yorker*, June 1, 2018, https://www.newyorker.com/science/elements/the-terrifying-lessons-of-a-pandemic-simulation.

11 Ibid.

12 Ibid.

13 World Health Organization, https://open.who.int/2018-19/contributors/contributor.

14 Natalie Huet, "Meet the world's most powerful doctor: Bill Gates," Politico.com, May 4, 2017.

15 "Dr. Cover-Up: Tedros Adhanom's controversial journey to the WHO," May 1, 2020, https://www.orfonline.org/expert-speak/dr-cover-up-tedros-adhanoms-controversial-journey-to-the-who-65493/.

Chapter Seven

1 "Mary Gates, 64; Helped her son start Microsoft," Associated Press, June 11, 1994.

2 Paul Allen, *Idea Man* (New York: Portfolio, 2011).

3 Pallava Bagla, "Indian Parliament comes down hard on cervical cancer trial," *Science*, September 9, 2013.

4 Ibid.

5 Aditya Kalra, "India Cuts Some Ties with the Gates Foundation on Immunization," Reuters, February 8, 2017.

6 Ibid.

7 https://www.gatesfoundation.org/How-We-Work/Quick-Links/Grants -Database/Grants/2020/03/INV-005273.

8 Amy Goodman, "Gates Foundation Causing Harm with the Same Money It Uses to Do Good," Democracy Now, January 9, 2007, https://www.democracynow.org/2007/1/9/report_gates_foundation _causing_harm_with.

9 Vandana Shiva, "Earth Democracy: Connecting Rights of Mother Earth to Human Rights and Well-Being of All," October 15, 2015, https://navdanyainternational.org/publications/earth-democracy -connecting-rights-of-mother-earth-to-human-rights-and-the-well -being-of-all/.

10 Dr. Joseph Mercola, "Vandana Shiva: Bill Gates Empires 'Must Be Dismantled,'" Children's Health Defense, March 29, 2021, https://childrenshealthdefense.org/defender/vandana-shiva-gates -empires-must-dismantle/.

11 Alex Ralph, "Bill Gates and George Soros buy out UK Covid test company Mologic," *The Times*, July 20, 2021, https://www.thetimes .co.uk/article/bill-gates-and-george-soros-buy-out-uk-covid-test-company -mologic-70c3r736b.

12 Emily Flitter and James B. Stewart, "Bill Gates Met With Jeffrey Epstein Many Times, Despite His Past," October 12, 2019, https://www.nytimes.com/2019/10/12/business/jeffrey-epstein-bill -gates.html.

Chapter Nine

1 https://casetext.com/statute/revised-code-of-washington/title-70-public-health-and-safety/chapter-70290-washington-vaccine-association/section-70290010-definitions.

2 Planet Lockdown Interview Series by James Henry, 2021, https://planetlockdownfilm.com.

3 Real America's Voice Radio, July 26, 2021, https://americasvoice.news/video/oLGAsJHJgdKwQPm/.

4 World Health Organization. "Global efforts to study the origin of SARS-CoV19 virus," August 2, 2020.

5 US Department of State Fact Sheet, "Activity at the Wuhan Institute of Virology," January 15, 2021, https://2017-2021.state.gov/fact-sheet-activity-at-the-wuhan-institute-of-virology/index.html.

Epilogue

1 Lauren Leatherby, Arielle Ray, Anjali Singhvi, Christiaan Triebert, Derek Watkins, Haley Willis, "Insurrection at the Capitol: A Timeline of How It Happened," *New York Times*, January 12, 2021, https://www.nytimes.com/interactive/2021/01/12/us/capitol-mob-timeline.html.

Be who you are and say what you feel.
Cuz those who mind don't matter and those who matter don't mind.

—Dr. Seuss